DOUBLE HELIX OF
PHYLLOTAXIS

DOUBLE HELIX OF PHYLLOTAXIS

ANALYSIS OF THE GEOMETRIC MODEL OF PLANT MORPHOGENESIS

BORIS ROZIN

BrownWalker Press
Irvine • Boca Raton

Double Helix of Phyllotaxis:
Analysis of the geometric model of plant morphogenesis

Copyright © 2020 Boris Rozin. All rights reserved.
No part of this publication may be reproduced, distributed, or transmitted in any form or by any means, including photocopying, recording, or other electronic or mechanical methods, without the prior written permission of the publisher, except in the case of brief quotations embodied in critical reviews and certain other noncommercial uses permitted by copyright law.

For permission to photocopy or use material electronically from this work, please access www.copyright.com or contact the Copyright Clearance Center, Inc. (CCC) at 978-750-8400. CCC is a not-for-profit organization that provides licenses and registration for a variety of users. For organizations that have been granted a photocopy license by the CCC, a separate system of payments has been arranged.

BrownWalker Press / Universal Publishers, Inc.
Irvine • Boca Raton
USA • 2020
www.BrownWalkerPress.com

978-1-62734-748-8 (pbk.)
978-1-59942-604-4 (ebk.)

Typeset by Medlar Publishing Solutions Pvt Ltd, India
Cover design by Ivan Popov

US Library of Congress Cataloging-in-Publication Data
is available at https://lccn.loc.gov

Names: Rozin, Boris, 1964- author.
Title: Double helix of phyllotaxis : analysis of the geometric model of plant morphogenesis / Boris Rozin.
Description: Irvine. California : Brown Walker Press, [2020] | Includes bibliographical references.
Identifiers: LCCN 2020019498 (print) | LCCN 2020019499 (ebook) | ISBN 9781627347488 (paperback) | ISBN 9781599426044 (ebook)
Subjects: LCSH: Phyllotaxis--Morphogenesis. | Plant morphogenesis--Mathematical models.
Classification: LCC QK649 .R855 2020 (print) | LCC QK649 (ebook) | DDC 581.4/8--dc23
LC record available at https://lccn.loc.gov/2020019498
LC ebook record available at https://lccn.loc.gov/2020019499

SYNOPSIS

This book is devoted to anyone who is in search of beauty in mathematics, and mathematics in the beauty around us. Attempting to combine mathematical rigor and magnificence of the visual perception, the author is presenting the mathematical study of phyllotaxis, the most beautiful phenomenon of the living nature. The distinctive feature of this book is an animation feature that explains the work of mathematical models and the transformation of 3D space.

The analysis of the phyllotactic pattern as a system of discrete objects together with the mathematical tools of generalized sequences made it possible to find a universal algorithm for calculating the divergence angle. In addition, it is serving as a new proof of the fundamental theorem of phyllotaxis and analytically confirming well-known formulas obtained intuitively earlier as well as casting some doubts on a few stereotypes existing in mathematical phyllotaxis.

The presentation of phyllotaxis morphogenesis as a recursive process allowed the author to formulate the hydraulic model of phyllotaxis morphogenesis and propose a method for its experimental verification. With the help of artificial intelligence, the author offered methodology for the digital measurement of phyllotaxis allowing a transition to a qualitatively new level in the study of plant morphogenesis. Due to the successful combination of mathematical constructions and their visual presentation, the materials of this study are comprehensible to readers with high school advanced mathematical levels.

CONTENTS

Acknowledgments		*xi*
I.	**INTRODUCTION**	**1**
II.	**MATHEMATICAL FOUNDATIONS**	**3**
	II.1. The Golden Ratio	3
	II.2. Recursion and recursive sequence	5
	II.2.1. The Fibonacci sequence	5
	II.2.2. Generalized recursive sequence	6
	II.2.3. The sum of recursive sequences	8
	II.3. Spirals	9
III.	**PHYLLOTAXIS**	**13**
	III.1. Phyllotaxis: concept and terminology	13
	III.2. Phyllotactic patterns classification and recursive sequence hierarchy	16
	III.3. Visual perception of phyllotactic patterns	18
	III.4. Double morphogenetic phenomenon of phyllotaxis	19
IV.	**THE DOUBLE HELIX MODEL OF PHYLLOTAXIS**	**21**
	IV.1. Geometric interpretation of the model	21
	IV.2. Bravais-Bravais theorem	24
	IV.3. The fundamental theorem of phyllotaxis	25
V.	**THE PLANAR DH-MODEL ON ARCHIMEDEAN SPIRALS**	**27**
	V.1. The criterion of the minimum distance between the nodes of the phyllotactic lattice	28
	V.2. Fibonacci phyllotaxis	29
	V.2.1. Finding the divergence coefficient for Fibonacci phyllotaxis	29
	V.2.2. The phyllotactic pattern with divergence coefficient $1/\tau^2$	31

	V.3.	The analysis of the minimum distances between the nodes in the Fibonacci phyllotactic lattice	32
	V.4.	The threads of the nodes of the phyllotactic lattice and their visual perception	33
	V.5.	Diapasons and nodes of "the phyllotaxis rises"	35
	V.6.	Diameter of an element of the phyllotactic lattice	39
	V.7.	The angle between tangent lines to the opposed parastichies	40
	V.8.	Is phyllotaxis a fractal or a recursive structure?	44
	V.9.	The planar DH-Model for "accessory" sequences	45
	V.10.	The planar DH-model for "lateral" sequences	49
	V.11.	Converse of the fundamental theorem of phyllotaxis	57
	V.12.	The planar DH-model for "multijugate" sequences	59
	V.13.	The universal algorithm for calculating the divergence coefficient	66
	V.14.	Interim results	67
VI.	**FIBONACCI LATTICES ON THE PLANAR NON-ARCHIMEDEAN SPIRALS**		69
	VI.1.	Fibonacci lattices on the planar power genetic spirals	70
	VI.2.	Fibonacci lattices on the planar logarithmic genetic spirals	74
	VI.3.	Interim results	77
VII.	**THE CYLINDRICAL DH-MODEL**		79
	VII.1.	The cylindrical Fibonacci phyllotaxis	82
	VII.2.	The cylindrical phyllotaxis generated by a non-multiple recurrent sequence	90
	VII.3.	The cylindrical phyllotaxis generated by the generalized recurrent sequence	92
VIII.	**PHYLLOTACTIC LATTICES WITH RATIONAL DIVERGENCE COEFFICIENT**		97
	VIII.1.	Phyllotactic lattices with the divergence coefficient $\beta = 0$	97
	VIII.2.	Phyllotactic lattices with the divergence coefficient $\beta = 0.5$	98
	VIII.3.	Two-pair phyllotactic lattices with the divergence coefficient $\beta = 0.5$	99
	VIII.4.	Multi-pair phyllotactic lattices with the divergence coefficient $\beta = 0.5$	100
	VIII.5.	Phyllotactic lattices with the divergence coefficient $\beta = 0.25$	101
	VIII.6.	The two-pair phyllotactic lattice with the divergence coefficient $\beta = 0.25$	102
	VIII.7.	Interim results	103

IX.	DISPUTE OF THE STATICS DH-MODEL	105
	IX.1. Novelty and validity of the DH-Model	105
	IX.2. Edge function and divergence coefficient	106
X.	SOME ASPECTS OF THE DYNAMIC DH-MODEL	109
	X.1. Phyllotaxis under the microscope or where is the inflorescence forming from?	109
	X.2. Invisible proprimordium and visible primordium	110
	X.3. The phenomenon of "cutting"	112
	X.4. The tube model of the periodic phyllotaxis	114
	X.5. The mathematical genesis of the Fibonacci numbers or why Turing had failed in his research in phyllotaxis	115
	X.6. The concept of the dynamic DH-Model	118
	X.7. The hydraulic aspect of the dynamic DH-Model	118
	X.8. Experimental verification of the hydraulic model	121
XI.	WHAT WE OBJECTIVELY KNOW AND DON'T KNOW ABOUT PHYLLOTAXIS	125
XII.	HOW PHYLLOTAXIS SHOULD BE MEASURED	127
	XII.1. Digital technology and Artificial Intelligence in the measurement of phyllotaxis	127
	XII.2. The network project "Open Phyllotaxis"	128
XIII.	CONCLUSION OR FIVE CRITICAL QUESTIONS	131

APPENDIX

A.	THE GOLDEN SECTION AND MORPHOGENESIS	135
	A.1. The Golden Ratio in the works of man	135
	A.2. The Golden Section in nature	136
	A.3. The Golden Section in technique	140
	A.4. Why does nature need the Golden Section?	140
B.	EXTENSIONS TO THE FIBONACCI SEQUENCE AND THE GOLDEN SECTION	143
	B.1. p-numbers Fibonacci	143
	B.2. Metallic Means	144
	B.3. Continuous Fibonacci functions	145

	B.4.	Bodnar's "golden" hyperbolic functions	150
	B.5.	The Golden sine	151
C.		THE FIBONACCI NUMBERS IN CYBERNETICS	155
D.		THE ANGLE OF THE TANGENT TO THE SPIRAL	157
E.		SOLVE A LINEAR DIOPHANTINE EQUATION	159
F.		TABLE OF THE DEPENDENCE OF THE DIVERGENCE COEFFICIENT AND ENCYCLIC NUMBERS ON THE INITIAL TERMS OF THE GENERATING RECURRENT SEQUENCE	161

References   165

Index   169

About Author   173

ACKNOWLEDGMENTS

The author is grateful to all researchers of the phenomenon of phyllotaxis. Biologists who have made thousands of observations of phyllotaxis patterns, accumulated and systematized a big array of information. To the mathematicians who tried to explain the numerical relationships in biological objects. Special thanks to Professor Roger V. Jean, for trying to systematize and summarize this knowledge.

The author thanks Intel and Microsoft, all the mathematicians and engineers who created modern computers and SW. Without using a computer with MS Visual Studio, MS WORD, and MS Excel, this study would have been impossible.

The author is grateful to his wife Natali and daughter Miriam for their help and support;

To Arkadiy Boshoer (Brooklyn, NY) for the criticisms and assistance in preparing this book;

To Dr. Alla Kolunuk (Kamianets-Podilskyi, Ukraine), to Valeria Boshoer (San Francisco, CA), to Robert Fuchs (München, Germany) for the help in preparation of this text;

To Dr. Jacques Dumais (Universidad Adolfo Ibáñez, Chile) for the kindly provided unpublished pictures and the criticisms.

INTRODUCTION

When I was kid and could not read, I thought philosophy and phyllotaxis had the same root.

Phyllotaxis, in many of its forms, has long been considered one of the most beautiful objects of the living nature. The first thing that catches your eye is the presence of a complex ordered structure, whether it be inflorescence, arrangement of leaves or seeds. The second, with same amount of curiosity, is the constant presence of integer ratios in the number of right- and left-handed spirals. And these are not just integers, but the Fibonacci numbers. The Fibonacci numbers are inextricably linked to the Golden Section, which is considered the canon (standard) of harmony and beauty.

There is a third feature of the phenomenon of phyllotaxis, it is an amazing consistency of a form. For example, imagine a walk in a pine forest, one would not imagine finding a single pair of pines with the same location and shape of branches. Without special attention, it is noticeable that all the trees are substantially different. However, on branches of various shapes, one spots absolutely identical cones! The same is noted about rose bushes: the branches are different, and the flowers are the same. This suggests that the morphogenetic processes for the inflorescences and the rest of the plants are somewhat different.

These studies answering the infamous childhood question "Why do we see spirals on the inflorescence of a sunflower?" This question seems very simple and has been answered long ago. However, an attempt to answer it unexpectedly raises a philosophical thought: "What are the visible spirals in a sunflower?", without an explanation of which it is impossible to answer a more deliberate question "Why is the number of left-handed and right-handed spirals equal to a pair of the Fibonacci numbers?" Indeed, without understanding what the object of counting is, it is impossible to understand nature of the phenomenon, or to know the morphological laws.

Unlike most of biological objects that are continuous systems and are described by statistical laws, phyllotaxis is a system of discrete objects that are mathematically described by stable integer relations.

The research of Bravais' brothers [1], published about 250 years ago, laid the foundation for mathematical phyllotaxis, the history of which is beautifully described in [2, 3]. However, despite the interest in this phenomenon from the best mathematicians of the nineteenth and twentieth centuries (Church [4], Turing [5, 6], Levitov [7], Adler [8], Coxeter [9]), no model was found that could explain the reason for the number of left-handed and right-handed spirals to be equal to a pair of the Fibonacci numbers. By analogy with other biosystems, most of researchers considered phyllotaxis as a continuous system using the mathematical apparatus of statistics, continuous functions, differential and integral calculus. However, **phyllotaxis** *is a collection of discrete elements, in mathematical description of which it is necessary to apply the methods of integer analysis.* Therefore, the reader will not find in this book mathematical calculations that go beyond the school course of mathematics. However, the reader will rather find previously unused mathematical tools of generalized recurrent series and an attempt to comprehend recursion as the basis of the morphology of the living nature.

The book contains online videos and images. All media files are highlighted in blue in the text (example: Video VI.1) and have HD resolution1080p. All videos and images are accessible on YouTube channel **Double Helix of Phyllotaxis** or by a direct link. If you use eBook (PDF file) direct link is clickable. Subscribers of YouTube channel Double Helix of Phyllotaxis will receive updates on the project "Open Phyllotaxis" and experiments to verification the hydraulic model.

MATHEMATICAL FOUNDATIONS

II.1. THE GOLDEN RATIO

The Golden Ratio (the Golden Section or the Golden Mean) is one of the most remarkable mathematical constants, along with π and e.

The classical definition of the Golden Section is the problem of proportional division of a line segment so that the ratio of a smaller part to a larger one, is the same the ratio of a larger to the whole segment

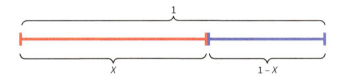

Figure II.1. Proportional division of a line in the Golden Ratio.

Let's denote the greater part of the unit interval X, then $\dfrac{1-X}{X} = \dfrac{X}{1}$ and find the roots of the quadratic equation

$$X^2 + X - 1 = 0 \qquad (II.1)$$

the positive root will be

$$X = \frac{-1+\sqrt{5}}{2} = 0.618..$$

Whereas the reciprocal value of $\dfrac{1}{X} = \dfrac{1+\sqrt{5}}{2} = 1.618....$ and usually denoted as τ and called the base of the Golden Ratio.

The main remarkable property of the Golden Ratio follows from (II.1)

$$\tau^2 = \tau^1 + 1$$

i.e. exponentiation is determined by addition, bypassing multiplication.

In this study, the ratios will be used:

$$\tau^{n+2} = \tau^{n+1} + \tau^n \tag{II.2}$$

$$1 + \frac{1}{\tau} = \tau$$

$$\tau + \frac{1}{\tau} = \sqrt{5}$$

Similarly to the problem of proportional division of a line segment, let's divide the circle into two arcs line (Figure II.2), where lengths are correlated as this is in Figure II.1. We obtain a smaller arc that equals to $\frac{2\pi}{\tau^2}$ and a larger one that equals to $\frac{2\pi}{\tau}$.

The smaller arc has an angle $\frac{2\pi}{\tau^2} \approx 137.5°$ and the larger one $\frac{2\pi}{\tau} \approx 222.5°$.

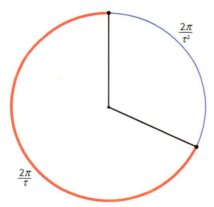

Figure II.2. Proportional division of a circle in the Golden Ratio.

If we unwrap the circle into a straight line (Figure II.3), we get a figure that is similar to Figure II.1.

Figure II.3. Proportional division of a unwrap circle in the Golden Ratio.

Accordingly, $\tau^2 = \tau + 1$ will look

$$2\pi = \frac{2\pi}{\tau} + \frac{2\pi}{\tau^2},$$

therefore:

$$\cos\left(\frac{2\pi}{\tau}M\right) = \cos\left(\frac{2\pi}{\tau^2}M\right),$$

$$\sin\left(\frac{2\pi}{\tau}M\right) = -\sin\left(\frac{2\pi}{\tau^2}M\right)$$

Appendix **A** presents a hypothesis explaining the presence of the Golden Ratio in nature. Appendix **B** presents the Golden Proportion extensions.

II.2. RECURSION AND RECURSIVE SEQUENCE

Recursion is the definition of the state of a system in terms of itself:

$$S_i = f(S_{i-1})$$

where S_i is the state of the system at the i-th moment of the time.

The most famous example of recursive systems is fractals. In mathematics, there are several integer functions that are also recursive, for example:

- The natural numbers $n = (n-1) + 1$
- Arithmetic progression $a_n = a_{n-1} + d$
- Factorial $n! = n \cdot (n-1)!$
- Exponential function $a^n = a \cdot a^{n-1}$, idem Geometric progression
- Various sequences $K_i = f(K_{i-1})$

II.2.1. The Fibonacci sequence

The most famous recurrent sequence is the Fibonacci sequence (the Fibonacci numbers or the Fibonacci series) 1, 2, 3, 5, 8, 13, 21, ..., whose members are calculated by the formula

$$F_{n+2} = F_{n+1} + F_n \qquad (II.3)$$

where $F_1=1$ and $F_2=2$ the initial terms of the sequence (Table II.2.1). Various sources give the Fibonacci sequence starting with $F_1=0$, $F_2=1$ or $F_1=1$, $F_2=1$, but it is the same sequence. In the literature, it is customary to denote F_n to the members of the Fibonacci sequence only.

The most interesting property of the Fibonacci numbers is

$$\lim_{n \to \infty} \frac{F_n}{F_{n-1}} = \tau = const$$

There is a well-known Binet's formula linking the Fibonacci sequence and the Golden Section:

$$F_n = \frac{\tau^{n+1} - (-1)^{n+1}\tau^{-(n+1)}}{\sqrt{5}} \qquad (II.4)$$

This formula is interesting because there is an integer on the left, whereas on the right, τ and $\sqrt{5}$ are irrational numbers. Binet's formula has many interesting forms reflecting its asymmetry.

$$F_n = \frac{\tau^{n+1} - \left(-\frac{1}{\tau}\right)^{n+1}}{\sqrt{5}} = \frac{\left(\frac{1+\sqrt{5}}{2}\right)^{n+1} - \left(\frac{1-\sqrt{5}}{2}\right)^{n+1}}{\sqrt{5}}.$$

For large n, it is possible to use formula for approximate calculation of the members of the Fibonacci sequence:

$$F_n \approx \frac{\tau^{n+1}}{\sqrt{5}}$$

Table II.2.1. Correspondence of n and n-th terms of the Fibonacci sequence.

n	1	2	3	4	5	6	7	8	9
F_n	1	2	3	5	8	13	21	34	55

II.2.2. Generalized recursive sequence

In the general case, the generalized recursive sequence can have any initial terms G_1, G_2 and is constructed by using the same recursive formula:

$$G_n = G_{n-1} + G_{n-2} \qquad (II.5)$$

Let us find a formula connecting the generalized recursive sequence (sometimes called the recursive sequence with arbitrary initial terms) with the Fibonacci sequence:

$$\begin{aligned}
G_3 &= & &= & G_2+G_1 \\
G_4 &= & G_3+G_2 &= & 2G_2+G_1 \\
G_5 &= & G_4+G_3 &= & 3G_2+2G_1 \\
G_6 &= & G_5+G_4 &= & 5G_2+3G_1 \\
G_7 &= & G_6+G_5 &= & 8G_2+5G_1 \\
G_8 &= & G_7+G_6 &= & 13G_2+8G_1 \\
&\vdots & &\vdots & \vdots \\
G_n &= & G_{n-1}+G_{n-2} &= & F_{n-2}G_2+F_{n-3}G_1
\end{aligned} \qquad (II.6)$$

Any recursive sequence is determined by the recurrent formula (II.5) and two initial terms. In terms of mathematics, two initial terms can be any two numbers, including negative, irrational, and even imaginary.

From (II.6) and (II.4) we can get Binet formula for the generalized recursive sequence [10]:

$$G_n = \frac{G_2+G_1\cdot\tau^{-1}}{\sqrt{5}}\tau^{n-1} - \frac{G_2-G_1\cdot\tau}{\sqrt{5}}(-\tau)^{-(n-1)} \qquad (II.7)$$

For the generalized recursive sequence:

$$\lim_{n\to\infty}\frac{G_n}{G_{n-1}} = \tau$$

For large n, we can use the formula for approximate calculations of the members of the generalized recursive sequence:

$$G_n \approx \frac{G_2+G_1\cdot\tau^{-1}}{\sqrt{5}}\tau^{n-1}$$

In this study, the following relationships between recursive sequences will be used:

$$\lim_{n\to\infty}\frac{F_{n-2}}{G_n} = \frac{\tau^{n-1}}{(G_2+G_1\cdot\tau^{-1})\tau^{n-1}} = \frac{1}{G_2+G_1\cdot\tau^{-1}}$$

$$\lim_{n\to\infty}\frac{F_{n-1}}{G_n} = \frac{\tau^n}{(G_2+G_1\cdot\tau^{-1})\tau^{n-1}} = \frac{\tau}{G_2+G_1\cdot\tau^{-1}}$$

$$\lim_{n\to\infty}\frac{H_n}{G_n} = \frac{(H_2+H_1\cdot\tau^{-1})\tau^{n-1}}{(G_2+G_1\cdot\tau^{-1})\tau^{n-1}} = \frac{H_2+H_1\cdot\tau^{-1}}{G_2+G_1\cdot\tau^{-1}} \qquad (II.8)$$

Where G_n and H_n are two different generalized recursive sequences, with initial terms, accordingly (G_1, G_2) and (H_1, H_2)

Also, this study will repeatedly use the recursive sequence with initial terms 1 and C_2, which we will call the Fibonacci-like sequence and denote, for clarity, as C_n,

$$1, C_2, C_2+1, 2C_2+1, 3C_2+2, 5C_2+3, 8C_2+5, \ldots, F_{n-2}C_2 + F_{n-3}, \ldots$$

$$\lim_{n\to\infty} \frac{F_{n-2}}{C_n} = \frac{1}{C_2 + \tau^{-1}} \qquad (II.9)$$

$$\lim_{n\to\infty} \frac{F_{n-1}}{C_n} = \frac{\tau}{C_2 + \tau^{-1}}$$

$$\lim_{n\to\infty} \frac{C_n}{G_n} = \frac{C_2 + \tau^{-1}}{G_2 + G_1 \cdot \tau^{-1}}$$

There are other extensions of the Fibonacci sequence, which are presented in Appendix **B**. Appendix **C** gives an overview of the application of the Fibonacci sequence and the Golden Ratio in cybernetics.

II.2.3. The sum of recursive sequences

By the sum of two recursive sequences we will further understand a recursive sequence, each term of which is equal to the sum of the corresponding terms of two sequences:

$$G_n = A_n + B_n \qquad (II.10)$$

The converse is also true: any recursive sequence can be represented as the sum of two recursive sequences, i.e. there is possible to do expansion any recursive sequence into a sum. In order to decompose the series G_n into a sum, it is sufficient to find the recursive sequence A_n.

The limit of the ratio of A_n and G_n sequences can be calculated using (II.8):

$$\lambda = \lim_{n\to\infty} \frac{A_n}{G_n} = \frac{A_2 + \dfrac{A_1}{\tau}}{G_2 + \dfrac{G_1}{\tau}} \qquad (II.11)$$

Accordingly, for B_n and G_n

$$1 - \lambda = \lim_{n\to\infty} \frac{B_n}{G_n} = \frac{B_2 + \dfrac{B_1}{\tau}}{G_2 + \dfrac{G_1}{\tau}} \qquad (II.12)$$

From (II.11) and (II.12), interesting relations follow:

$$A_n = [\lambda G_n] \approx \lambda G_n \tag{II.13}$$

$$B_n = [(1-\lambda)G_n] \approx (1-\lambda)G_n \tag{II.14}$$

where [ ] it is rounded to the nearest integer.

II.3. SPIRALS

Traditionally, there are two types of spirals in mathematics: planar and cylindrical.

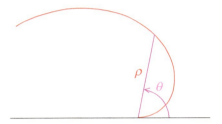

Figure II.4. A polar coordinate system.

In a polar coordinate system, a planar spiral is defined by the formula $\rho = E(\theta)$, where θ is the angle of rotation of the radius vector, ρ is the length of the radius vector (Figure II.4), $E(\theta)$ is a real continuous and monotonic increasing function, which will be called the Edge function.

In a Cartesian coordinate system, a planar spiral is defined by a parametric system of equations:

$$\begin{cases} x(\theta) = E(\theta) \cdot \cos(k\theta) \\ y(\theta) = E(\theta) \cdot \sin(k\theta) \end{cases} \tag{II.15}$$

where is k real constant.

There is no one-to-one correspondence between the representation of a planar spiral in polar and Cartesian coordinate systems. For example, the spiral $\rho = l\theta$ can be represented:

$$\begin{cases} x(\theta) = \cos(l\theta) \\ y(\theta) = \sin(l\theta) \end{cases} \quad \text{or} \quad \begin{cases} x(\theta) = l \cdot \cos(\theta) \\ y(\theta) = l \cdot \sin(\theta) \end{cases}$$

Therefore, a Cartesian coordinate system will be used in this research.

From the whole variety of Edge functions, one can distinguish the exponential $ae^{b\theta}$ and the power $a\theta^v$ functions.

If the Edge function is an exponent, then such a planar spiral is called a logarithmic spiral:

$$\begin{cases} x(\theta) = ae^{b\theta} \cdot \cos(k\theta) \\ y(\theta) = ae^{b\theta} \cdot \sin(k\theta) \end{cases}$$

where a, b, and k are real constants.

If the Edge function is a power function, then such a planar spiral will be called a power spiral:

$$\begin{cases} x(\theta) = a\theta^{\nu} \cdot \cos(k\theta) \\ y(\theta) = a\theta^{\nu} \cdot \sin(k\theta) \end{cases}$$

where a, ν, and k are real constants.

Among the power spirals, the following should be highlighted:

- when $\nu=1$, then there is the Archimedean spiral $E(\theta) = a\theta$, which is the simplest spiral;
- when $\nu=1/2$, then there is the parabolic spiral is $E(\theta) = a\theta^{\frac{1}{2}}$;
- when $\nu=1/\tau$, then there is the "golden" spiral is $E(\theta) = a\theta^{\frac{1}{\tau}}$.

Appendix **D** provides very useful formulas, in the study of phyllotaxis, finding the angle of inclination of the tangent to the spirals.

A helix (or cylindrical spiral) is a three-dimensional curve located on a straight circular cylinder. Usually, a cylindrical spiral is understood to mean a uniform helix, which is defined by a parametric system in three-dimensional space

$$\begin{cases} x(\theta) = R \cdot \cos(k\theta) \\ y(\theta) = R \cdot \sin(k\theta) \\ z(\theta) = H \cdot \theta \end{cases}$$

where R is the radius of the cylinder and H is the step of the helix.

In mathematics, there are known figures of the rotation the surface of which is formed by rotating a certain curve around the axis of the symmetry. A plane can be defined as the rotation of a straight line perpendicular to the axis of the rotation, and a cylinder as the rotation of a straight line parallel to the axis of the rotation. From this it follows that the plane and the cylinder are the two limited shapes of the figures of rotation.

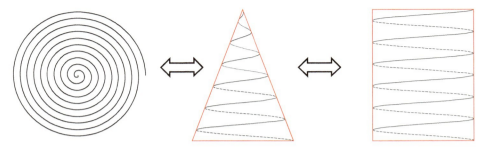

Figure II.5. Transformation the planar Archimedean spiral into a helix.

Therefore, it is possible to assert the existence of a non-linear transformation of space that transforms the planar Archimedean spiral into a helix one and vice versa (Figure II.5). Video_II.1[1] explains the transformation of the planar Archimedean spiral into a spiral on a hemisphere, then successively into a spiral on a cone, into a helix (or cylindrical spiral), into a spiral on a hyperboloid of revolution, and into a spiral on a figure of rotation similar to a Conifer cone.

We can say that the planar Archimedean spiral can be "pulled" onto any rotation surface, while the axis of the spiral and the axis of the rotation figure will coincide. The reverse is also true: any spiral "pulled" onto the rotation figure can be converted into an Archimedean spiral. Moreover, the properties of these spirals associated with the relative position of the points on these spirals will be preserved.

Another non-linear transformation of space is the projection of the Archimedean spiral on various figures of rotation. In Video_II.2[2], the logarithmic spiral is represented as a projection of the planar Archimedean spiral on a hyperbolic funnel, and in Video_II.3[3], the Power spiral is represented as a projection of an Archimedean spiral on a parabolic cone of rotation. In a case of such non-linear transformation of the space, the properties of the spirals associated with the mutual arrangement of points on the spirals will also remain.

These properties of spirals allow first, to carry out a detailed analysis of the model of phyllotaxis on the planar Archimedean spirals as on the simplest ones, and second, to expand the analysis for logarithmic, power, and cylindrical spirals.

---

[1] The video is located at https://youtu.be/PfB3jz_IWdI
[2] The video is located at https://youtu.be/XagPu0vE2GA
[3] The video is located at https://youtu.be/BnlwoNoHUFk

PHYLLOTAXIS

III.1. PHYLLOTAXIS: CONCEPT AND TERMINOLOGY

Phyllotaxis (from the Greek *phýllon* — leaf and *táxis* — arrangement, adjective *phyllotactic*), covers a very wide range of botanical objects, in the structure of which orderliness, helicity, periodicity or symmetry are observed. These structures, which are often surprisingly beautiful, are usually called *phyllotactic patterns.* Phyllotactic patterns consist of discrete elements, such as new shoots, leaves, flower petals, and seeds, which are called *primordium* (plural *primordia*, adjective *primordial*).

For researchers, plant objects with a complex, pronounced spiral structure consisting of a large number of primordia are of particular interest. In this study, by phyllotaxis we will understand spiral phyllotaxis, unless otherwise specified.

On these objects, right- and left-handed spirals, which are called *parastichy*, are clearly distinguishable. It is these spirals that attract researchers with their beauty and mysteriousness.

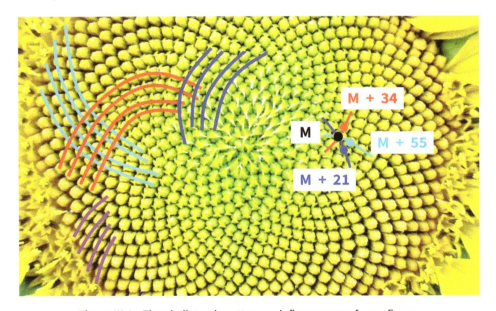

Figure III.1. The phyllotactic pattern on inflorescence of a sunflower.

In Figure III.1, a photograph of the classic planar phyllotactic pattern on inflorescence of a sunflower. In this pattern, four parastichies families are clearly visible — two parastichies families are twisted counterclockwise (cyan and blue) and two clockwise (violet and red), with the number of parastichies in each family being equal to one of the Fibonacci numbers: 89 parastichies in the "violet" family, 34 parastichies in the "red" family, 55 in the "cyan" family, and 21 in the "blue" family. The family containing F_n of various parastichies is usually called *the n-th family of parastichies.*

Usually on the phyllotactic pattern, a person can clearly distinguish at least two parastichies twisted in different directions, which are called *the opposed parastichy pair*, and a pair of numbers (F_n, F_{n+1}) which are called the *parastichy index*, where F_n is the number of parastichies in the *n*-th family. For example, a pair of "red" and "cyan" parastichies has an index (34, 55), a pair of "blue" and "red" has an index (21, 34). In Figure III.1, you can see the transition of one pair of opposing parastichies (index (21, 34)) to another pair (index (34, 55)), this is commonly called *"the phyllotaxis rises."*

In the overwhelming majority (from 87% to 97% according to Jean [3]), the number of spirals in a family is equal to one of the Fibonacci numbers, the so-called Fibonacci phyllotaxis or Fibonacci pattern. There are also non-Fibonacci patterns in which the number of spirals on the right and left in the family of parastichies is equal to the neighboring members of the generalized recurrent sequence. Such a generalized sequence will be called a generating sequence for this pattern. And, although the probability of the appearance of non-Fibonacci patterns is small, their analysis will help to build an authentic model of phyllotaxis.

In addition to the planar pattern, there are also distinguished phyllotactic patterns, in which primordia are located on 3D rotation figures, primarily on a cylinder.

Figure III.2. Visible parastichies on a spruce cone **A** (background picture from FreeImages.com/Michal Zacharzewski), pinecone **B**, pineapple **C**.

The classic examples of phyllotaxis on a cylinder is the pineapple fruit (Figure III.2.C) and the fir cone (Figure III.2.A). In a cylindrical phyllotaxis, parastichy is a collection of primordia located on a helix. As it was mentioned above, the number of parastichies in one family is usually equal to the Fibonacci number.

At the base of the pinecone (Figure III.2.B), the phyllotactic pattern is very close to the planar phyllotaxis, then the parastichies smoothly passes to the cylindrical phyllotaxis. The daisy flower pattern (Figure III.2.C) has the shape of a cone or paraboloid of rotation.

The spiral aloe phyllotactic pattern (Figure III.3.B) has only one clearly visible family of parastichy. The artichoke fruit phyllotactic pattern (Figure III.3.A) and the rose flower (Figure III.3.D) do not have any clearly visible parastichies, but they have a clearly visible spiral structure. The spiral structure is due to the primordia, where the petals of the artichoke or the rose, are arranged on a spiral, and their size increases from the center of the spiral to the outer edge. On the Internet, you can find many photos of various phyllotactic patterns, for example [11].

Figure III.3. Spiral and volumetric phyllotaxis. Artichoke **A**, Spiral Aloe **B**, Daisy **C** (picture from FreeImages.com/Emin Ozkan), Rose **D**.

However, simple structures are the most common in nature and they are the arrangements of the leaves, which are perceived by the observer as periodic structures. Figure III.4 shows examples of such structures, which have a periodically repeated arrangement of primordia, but do not have visible spirals in the patterns. Because of their simplicity, such periodical patterns of phyllotaxis practically do not attract the attention of researchers.

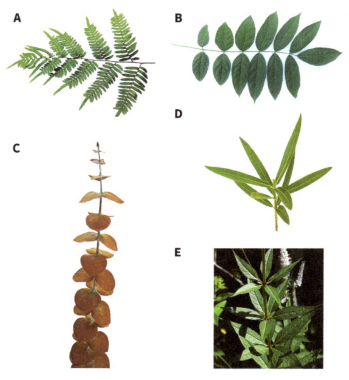

Figure III.4. Periodical phyllotaxis: distichous or alternate (**A**), opposite superposed (**B**), opposite decussate (**C**), 3–whorled phyllotaxis (**D**), 5–whorled phyllotaxis (**E**).

III.2. PHYLLOTACTIC PATTERNS CLASSIFICATION AND RECURSIVE SEQUENCE HIERARCHY

Jean in [3] proposed the classification of phyllotactic patterns:

"Main" sequence (the Fibonacci sequence) $F_1 = 1$ and $F_2 = 2$ 1, 2, 3, 5, 8, 13, 21,
"Accessory" sequences:
    First "accessory" sequence   1, 3, 4, 7, 11, 18 … (Lucas numbers),
    Second "accessory" sequence   1, 4, 5, 9, 14, 23 …,
    Third "accessory" sequence   1, 5, 6, 11, 17, 28 …,

"Multijugate" sequence:
    "Bijugate" main sequence   2, 4, 6, 10, 16, 26, ...
    "Bijugate" First accessory sequence   2, 6, 8, 14, 22, 36, ...
    "Bijugate" second accessory sequence   2, 8, 10, 18, 28, 44, ...
    "Trijugate" main sequence   3, 6, 9, 15, 24, 39, ...
    "Quadrijugate" main sequence   4, 8, 12, 20, 32, 52, ...
"Lateral" sequence or "Anomalous" phyllotaxis:
    First "lateral" sequence: 2, 5, 7, 12, 19, ...
    Second "lateral" sequence: 2, 7, 9, 16, 25, ...
    Third "lateral" sequence: 2, 9, 11, 20, 31, ...

In this research, the terms "main", "accessory", "multijugate", "lateral", and "anomalous" are all written in quotation marks to indicate the conditional aspect of these terms.

In mathematical phyllotaxis, only integer recurrent sequences are considered, because the subject of the research is the relative position of discrete objects — primordia. However, the same recurrent sequence can be defined by different pairs of initial terms. For example, the sequences with initial terms ($G_1 = 4$, $G_2 = 7$) and ($G_1 = 18$, $G_2 = 29$) — are the same sequences with initial members 1 and 3. Therefore, it is necessary to standardize the generating recurrent sequences so that each sequence will be determined by only one pair of initial terms. The author introduces restrictions on the initial members of the generating recurrent sequences:

$$\begin{cases} G_1, G_2 \in N \\ 1 \leq G_1 \\ 2G_1 \leq G_2 \end{cases} \quad (III.1)$$

As a result of these restrictions, the Fibonacci sequence is defined as a recurrent sequence with the initial members 1 and 2. From the restriction (III.1), it follows that the Fibonacci sequence has the least initial terms among all the generating sequences. The generating recurrent sequences can be placed in the following hierarchy:

The Fibonacci sequence ($F_1 = 1$ and $F_2 = 2$):

$$1, 2, 3, 5, 8, 13, 21, ...$$

The Fibonacci-like sequence ($C_1 = 1$):

$$1, C_2, C_2+1, 2C_2+1, 3C_2+2, 5C_2+3, ..., F_{n-2}G_2 + F_{n-3}, ...$$

Non-multiple recurrent sequence (the first two members of the series are coprime numbers, gcd(G2, G1) = 1):

$$G_1, G_2, G_2+G_1, 2G_2+G_1, 3G_2+2G_1, 5G_2+3G_1, ..., F_{n-2}G_2 + F_{n-3}G_1, ...$$

Generalized recurrent sequence ($j \geq 1, j \in \mathbb{N}$, $\gcd(G_2, G_1) = 1$):

$$jG_1, jG_2, jG_2 + jG_1, 2jG_2 + jG_1, 3jG_2 + 2jG_1, 5jG_2 + 3jG_1, \ldots, jF_{n-2}G_2 + jF_{n-3}G_1, \ldots$$

As it can be seen from the above hierarchy of recursive sequences, each group of sequences is a special case of the previous group. If you rearrange points 3 and 4 to the list of phyllotactic patterns from [3], then it is not difficult to see the full correspondence between the hierarchy of the generating recursive sequences and the classification of phyllotactic patterns.

Although 90% of the observed patterns of phyllotaxis have Fibonacci phyllotaxis, it is impossible to understand the phenomenon of phyllotaxis without a close analysis of non-Fibonacci phyllotaxis.

III.3. VISUAL PERCEPTION OF PHYLLOTACTIC PATTERNS

Jean [3] has repeatedly noted the problem of recognizing phyllotactic patterns, which consists in the absence of strict formal signs of belonging of some primordium to some parastichy. According to the author, this is a systemic problem, due to the fact that the "measuring instrument" of phyllotaxis is human perception. The human sense organs are imperfect; it is not so difficult to deceive our perception. Just look at the Escher's pictures [12] or installation of Bernard Bras [13]. Such visual effects are called optical illusions — the main source of which is the unconscious correction of visual images. The most routine optical illusion happens when we combine images, at a frequency of 24 frames per second, as a result of it, we can see a changing, moving image.

Before starting to do a quantitative analysis, it is necessary to answer the question: what do we actually see when looking at the phyllotactic pattern? At first glance, this question seems to be rather simple, even childish, but as it will be shown later, the formalization of what we see will allow the analysis to go further than the previous researchers reached.

Let's look at inflorescence of a sunflower (Figure III.1), which is a classic example of the phyllotaxis pattern. The first thing that catches the eye is the right- and left-handed spirals, which are formed by primordia of the inflorescence of a sunflower (in the case of a sunflower, primordium is a seed). A closer look at these spirals reveals two features:

- the number of identical spirals is equal to one of the Fibonacci numbers;
- the spirals are not infinite. When looking at it closer, we can notice the initial and final seeds between which this spiral is visible. Moreover, the initial seeds of these spirals are not in the center of the inflorescence, and the final ones may not be on the outer border.

In this Section, it is necessary to turn to biology and answer the question: "How does a seed differ from the other parts of a plant?" The roots, stem, branches, and other parts of a plant perform their function to ensure the vital activity of the whole plant. On the contrary, the function of a seed is to be *separated*. A seed is a part of a plant that will be separated from the parent plant in order to give rise to a new plant, i.e. each seed is a discrete object that will be separated from the plant that generates it.

If a seed is a discrete object, then what are the spirals that we see? Our brain, unconsciously, integrates discrete elements into pseudo-objects that look like spirals. The spirals that we see on the phyllotactic pattern exist only in our brain.

Phyllotaxis is not only one visual effect which combines homogeneous discrete objects into a pseudo-object. The most vivid example of such an effect is the merging of the stars into constellations, which are unified objects only in human perception. It is known from the school course of astronomy that stars and galaxies, which are combined in a constellation, are visually located at a small angular distance. However, the linear distances between the stars, in the same constellation, can be hundreds and thousands of light years, but to the terrestrial observer, these stars seem to be "near." To be fair, it should be noted that the astronomy inherited constellations from the pseudoscience astrology.

In general, for our visual perception to unite homogeneous discrete objects, they do not need to have any physical or logical interconnection; it is enough for the observer to see them under a minimum angular distance.

To exclude subjective human perception from the analysis of the phenomenon of phyllotaxis, the following statements must be accepted:

– phyllotaxis pattern is a system of discrete elements (primordia);
– human perception unites primordia, being at the minimum distance from each other, into spirals (parastichies);
– spirals on the phyllotactic pattern are a visual effect;
– phyllotaxis is a visual perception of the spatial structure of primordia.

This will allow move us to an objective analysis, the main object of which is the structure of the mutual arrangement of primordia, but not visible parastichies on the phyllotactic pattern.

III.4. DOUBLE MORPHOGENETIC PHENOMENON OF PHYLLOTAXIS

Morphogenesis (from the Greek μορφή — shape and γένεσις — creation), literally, is "forming of the shape." In the most modern scientific works, the term morphogenesis is mainly used in the sense laid down by Alan Turing in his classic work

"The Chemical Basis of Morphogenesis" [14] to describe the processes at the molecular and cellular levels. In this research, the term morphogenesis will be used to refer to a wider range of processes that describe forming of the shape a biological object. Accordingly, a mathematical model of morphogenesis should describe why one or another object has just such a shape and internal structure.

As it was shown in Section III.3, from the point of view of morphogenesis, a seed is a discrete object (primordium), whereas the plant that gave rise to primordium is an object having a complex structure — roots, stem, branches, etc. One of the characteristics of a plant, as an object and structure, is the continuity of form in space.

Then, from the point of view of morphogenesis, phyllotaxis has a double phenomenon:

- a continuous object generates discrete objects;
- these discrete objects form an ordered structure.

Such an understanding of phyllotaxis allows the author to assume that the analysis of the morphogenesis of phyllotaxis will provide a key to understanding the process of all biological forming of the shape.

THE DOUBLE HELIX MODEL OF PHYLLOTAXIS

IV.1. GEOMETRIC INTERPRETATION OF THE MODEL

Based on the work of many researchers, Bravais brothers [1], Church [4], Jean [15, 3], and Adler [8], it is possible to formulate the geometric model of phyllotaxis as a mathematical abstraction that allows one to analyze the phenomenon of phyllotaxis. This geometric model assumes that there are two oppositely twisted spirals (or helixes):

$$\rho_m(\theta) = E(\theta) \cdot (2\pi\beta\theta) \qquad (IV.1)$$

$$\rho_a(\theta) = -E(\theta) \cdot (2\pi(1-\beta)\theta) \qquad (IV.2)$$

where $\theta \geq 0$ is an independent real variable, $E(\theta)$ is a real continuous and monotonic increasing function which will be called the Edge function, and $0 \leq \beta \leq 0.5$ is a real coefficient, the meaning of which will be determined later. The spiral (IV.1) we will call *the main genetic spiral* and the (IV.2) *the ancillary genetic spiral*.

The main and the ancillary genetic spirals intersect at points, which will be called *the nodes of the phyllotactic lattice* or simply nodes. Accordingly, the mutual disposition of nodes is called the phyllotactic lattice. The phyllotactic lattice node is a mathematical abstraction of a biological object of primordium, and the phyllotactic lattice is an abstraction of the phyllotactic pattern.

In mathematics, we distinguish two boundary types of spirals: planar and cylindrical. In phyllotaxis, we also distinguish two boundary forms: planar phyllotaxis and cylindrical phyllotaxis.

For planar phyllotaxis, the genetic spirals in a Cartesian coordinate system are described by a parametric system:

$$\begin{cases} x_m(\theta) = E(\theta) \cdot \cos(2\pi\beta\theta) \\ y_m(\theta) = E(\theta) \cdot \sin(2\pi\beta\theta) \end{cases} \qquad (IV.3)$$

$$\begin{cases} x_a(\theta) = E(\theta) \cdot \cos(-2\pi(1-\beta)\theta) \\ y_a(\theta) = E(\theta) \cdot \sin(-2\pi(1-\beta)\theta) \end{cases} \quad (IV.4)$$

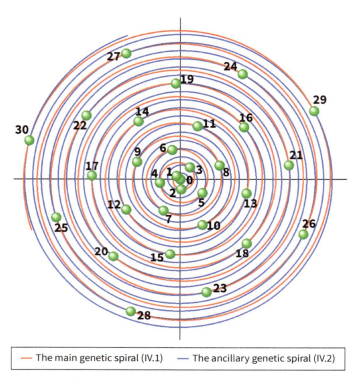

— The main genetic spiral (IV.1)   — The ancillary genetic spiral (IV.2)

Figure IV.1. The planar phyllotactic lattice.

Spirals (IV.3) and (IV.4) have a single center and intersect at points with an integer value θ, which we called *nodes*. The θ value in the intersect point will be called the number M ($M \in N$) of the node of the phyllotaxis lattice (or simply *the node M*). These nodes will have the coordinates in a Cartesian coordinate system:

$$\begin{cases} x(M) = E(M) \cdot \cos(2\pi\beta M) \\ y(M) = E(M) \cdot \sin(2\pi\beta M) \end{cases} \quad (IV.5)$$

The angle of rotation of the radius vector sliding along the main genetic spiral from one node to the next is $2\pi\beta$. And the angle of rotation of the radius-vector, sliding along the ancillary genetic spiral from the same node to the next is $2\pi(1-\beta)$.

The angle $2\pi\beta$ is usually called the divergence angle, and β is *the divergence coefficient*. Later, we will use the term divergence coefficient, since it matches the essence of the model more closely.

In research works on phyllotaxis, one of the main numerical parameters describing the phyllotaxis pattern is the *Plastochrone ratio R*. Jean [3] defines the "Plastochrone ratio R" as the ratio of the distances from the centers of two successive primordia to the center of the inflorescence. As noted above, the center of each primordium corresponds to the phyllotactic lattice site. Then "Plastochrone ratio R" will be:

$$R = \frac{E(M+1)}{E(M)}$$

or

$$\log(R) = \log(E(M+1)) - \log(E(M))$$

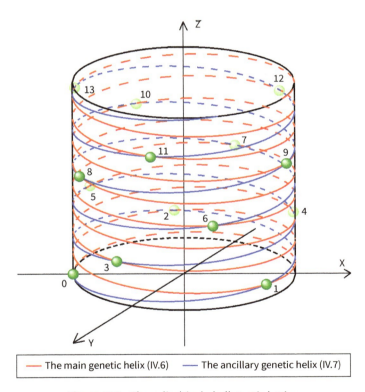

Figure IV.2. The cylindrical phyllotactic lattice.

In cylindrical phyllotaxis, genetic spirals are helixes that lie on the surface of a cylinder:

$$\begin{cases} x_m(\theta) = R \cdot \cos(2\pi\beta\theta) \\ y_m(\theta) = R \cdot \sin(2\pi\beta\theta), \\ z_m(\theta) = H_m \beta\theta \end{cases} \quad \text{(IV.6)}$$

$$\begin{cases} x_a(\theta) = R \cdot \cos(-2\pi(1-\beta)\theta) \\ y_a(\theta) = R \cdot \sin(-2\pi(1-\beta)\theta), \\ z_a(\theta) = H_a(1-\beta)\theta \end{cases} \quad \text{(IV.7)}$$

where R is the cylinder radius, H_m is the step of the main genetic helix, H_a is the step of the ancillary genetic helix.

H_m and H_a are interconnected by the formula:

$$\frac{H_m}{H_a} = \frac{\beta}{1-\beta}$$

Therefore, in the future, we will denote the step of the main genetic helix as H. Nodes of the phyllotactic lattice will have the coordinates:

$$\begin{cases} x(M) = R \cdot \cos(2\pi\beta M) \\ y(M) = R \cdot \sin(2\pi\beta M) \\ z(M) = H \cdot \beta M \end{cases}$$

For the sake of brevity, we will call the model of phyllotaxis, which is based on two oppositely twisted spirals or helices, as the *DH-Model* (DH – double helix).

IV.2. BRAVAIS-BRAVAIS THEOREM

The assumption of the Bravais brothers is that one spiral can be drawn through all the primordia of the phyllotactic pattern was the most important step in the development of mathematical phyllotaxis. According to the Bravias' [1], if all primordia lie on one spiral, then it is possible to number the primordia by the distance from the beginning of the spiral. Next, choose an arbitrary primordium with the number M and the n-th parastichy family. It is assumed that one parastichy of the n-th family, in its visible part, passes through primordium M. Since the primordia are numbered from the center of the spiral, the primordium with the number $M + 1$ will be farther from the center than the primordium with the number M and belong to another parastichy from the same n-th family. Each next primordium will be located further away from the center and belong to another parastichy from the same n-th family. However, the primordium with the number $M + F_n$ will belong to the same parastichy as the primordium M, because, in this n-th family, there are only

F_n parastichies. For primordium with number M, primordium with number $M + F_n$ will be nearby, i.e. the distance between them will be minimal (Figure III.1).

The statement that "the numbers of two neighboring primordia lying on the same parastichy of the n-th family differ by F_n" is called the Bravais-Bravais theorem. It should be noted that the Bravais-Bravais theorem is valid for other patterns, not only for the Fibonacci phyllotaxis. If the phyllotactic pattern is generated by a recurrent sequence with initial terms G_1 and G_2, then the n-th family will have different G_n parastichies and the numbers of nearby primordia, lying on the same parastichies of this family, will differ on G_n.

IV.3. THE FUNDAMENTAL THEOREM OF PHYLLOTAXIS

The fundamental meaning of this theorem is indicated by its name. Jean in [3, 15] formulates the fundamental theorem of phyllotaxis as:

"Let (m, k) be a parastichy pair, where m and k are relatively prime, in a system with divergence angle d. The following properties are equivalent:

(1) There exist unique integers $0 \leq v < k$, and $0 \leq u < m$ such that

$$|mv - ku| = 1,$$

and $d \leq \frac{1}{2}$ is in the closed interval whose endpoints are u/m and v/k;

(2) The parastichy pair (m, k) is visible and opposed."

It should be noted that at the angle of divergence d, Jean [3] meant the divergence coefficient because the angle of divergence lies in the interval $\left[\frac{2\pi u}{m}, \frac{2\pi v}{k}\right]$.

This theorem connects the index of opposed parastichies (m, k) and the divergence angle using a "mystery" u and v, which Jean [3] called encyclic numbers. In this study, the meaning and method (algorithm) of calculating these numbers will be found.

THE PLANAR DH-MODEL ON ARCHIMEDEAN SPIRALS

As it is known in mathematics, planar uniform spirals are the simplest; by tradition they are called Archimedean spirals. Therefore, for the analysis of the DH-Model, we will begin with the Archimedean spirals.

If in the DH-Model the genetic spirals are Archimedean, then $E(\theta) = \theta$:

$$\begin{cases} x_m(\theta) = \theta \cdot \cos(2\pi\beta\theta) \\ y_m(\theta) = \theta \cdot \sin(2\pi\beta\theta) \end{cases} \quad (V.1)$$

$$\begin{cases} x_a(\theta) = \theta \cdot \cos(-2\pi(1-\beta)\theta) \\ y_a(\theta) = \theta \cdot \sin(-2\pi(1-\beta)\theta) \end{cases} \quad (V.2)$$

For brevity, we will call such *the Archimedean phyllotactic lattice*, or a phyllotaxis on Archimedean spirals.

The node M of the Archimedean phyllotactic lattice will have the following coordinates:

$$\begin{cases} x(M) = M \cdot \cos(2\pi\beta M) \\ y(M) = M \cdot \sin(2\pi\beta M) \end{cases} \quad (V.3)$$

Let's find the "Plastochrone ratio R" for the phyllotaxis on Archimedean spirals:

$$R = \frac{M+1}{M} = 1 + \frac{1}{M} \quad (V.4)$$

or

$$\log(R) = \log\left(\frac{M+1}{M}\right) = \log(M+1) - \log(M)$$

As it is known in mathematics, the decomposition is known as Taylor's Row.

$$\log\left(\frac{x+1}{x}\right) = 2\left(\frac{1}{2x+1} + \frac{1}{3(2x+1)} + \frac{1}{5(2x+1)} + \ldots\right)$$

This allows us to calculate the approximate value of the "Plastochrone ratio R" for large values of M:

$$\log(R) = \log\left(\frac{M+1}{M}\right) \approx \frac{1}{M+\frac{1}{2}} \quad (V.5)$$

As it is seen from (V.4) and (V.5), R or log (R) is not a constant or linear function from M and does not depend on the divergence coefficient for the Archimedean phyllotaxis.

V.1. THE CRITERION OF THE MINIMUM DISTANCE BETWEEN THE NODES OF THE PHYLLOTACTIC LATTICE

As it was discussed in Section IV.1, visible parastichies are the visual effect of combining primordia in a spiral. The author suggests that our brain combines primordia, which are at a minimum distance from each other. Therefore, at the first stage, the analysis of the phyllotactic lattice is viewed as a search of mathematical constants of the relative position of the nearby nodes.

Let's analyze two arbitrary lattice sites of phyllotaxis with numbers M and K ($M, K \in N$) and the center of the spiral O.

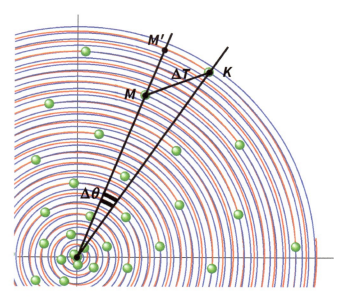

Figure V.1. Nodes M and N at the planar phyllotactic lattice.

The lengths of the segments OM and OK in accordance with (V.3) are equal to M and K. Let's mark the length of the segment between nodes M and K as ΔT (Figure V.1). According to the Cosine theorem:

$$\Delta T^2 = M^2 + K^2 - 2MK \cdot \cos(\Delta\theta)$$

Subtract and add $2MK$:

$$\Delta T^2 = M^2 - 2MK + K^2 - 2MK \cdot \cos(\Delta\theta) + 2MK = (M-K)^2 + 2MK(1-\cos(\Delta\theta))$$

Use the half-angle formula $1 - \cos(\gamma) = 2\sin^2(\gamma/2)$:

$$\Delta T^2 = (M-K)^2 + 4MK \sin^2(\Delta\theta/2) \tag{V.6}$$

As it can be seen from (V.6), both terms on the right-hand side of the equality are positive for any M, K, and $\Delta\theta$. In order for ΔT to be minimal, each of the terms on the right-hand side of the equality must tend to the minimum:

$$\begin{cases} |M-K| \to \min \\ |\Delta\theta| \to 0 \end{cases}$$

V.2. FIBONACCI PHYLLOTAXIS

Fibonacci phyllotaxis is the most common and researched type of phyllotaxis. Numerous sources indicate that Fibonacci phyllotaxis has a divergence coefficient $\dfrac{1}{\tau^2}$.

To prove the functional completeness of this model, it is necessary to prove the identical relationship between the divergence coefficient and the number of observed right- and left-handed spirals in the phyllotactic pattern, i.e. both statements must be true:

- Fibonacci phyllotaxis is possible only with a divergence coefficient $\dfrac{1}{\tau^2}$;
- with a divergence coefficient $\dfrac{1}{\tau^2}$, there is possible Fibonacci phyllotaxis only.

Let us prove each statement independently.

V.2.1. Finding the divergence coefficient for Fibonacci phyllotaxis

According to the Bravais-Bravais theorem, in Fibonacci phyllotaxis, primordia with numbers M and $M + F_n$ belong to the same parastichy and are close to each other.

Let's find the angle by which the radius-vector rotates from the node M to the node $M + F_n$, sliding along the main genetic spiral:

$$\Theta_m = 2\pi\beta((M+F_n) - M) = 2\pi\beta F_n$$

If the radius-vector slides along an ancillary genetic spiral, then the angle will be

$$\Theta_a = -2\pi(1-\beta)(M - (M+F_n)) = 2\pi(1-\beta)F_n$$

The number of complete turns of the radius-vector from the node M to the point M' (Figure V.1) along the main genetic spiral is indicated by A_n, and along the ancillary genetic spiral is B_n. The numbers A_n and B_n are integers because they figure out the number of complete rotations around the center. Therefore, the coefficient $|\Delta\theta|$ can be calculated in two ways:

$$|\Delta\theta| = |\Delta\Theta_m - 2\pi A_n| = 2\pi|\beta F_n - A_n| \tag{V.7}$$

$$|\Delta\theta| = |2\pi B_n - \Delta\Theta_a| = 2\pi|B_n - (1-\beta)F_n|$$

Because $|\Delta\theta| \to 0$, then

$$A_n \approx \beta F_n \tag{V.8}$$

$$B_n \approx (1-\beta)F_n \tag{V.9}$$

As it is shown in Sections II.2.3, A_n and B_n are members of some integral recurrent series. Moreover, $A_n < B_n$ because $0 < \beta < 0.5$. If we sum (V.8) and (V.9) we get

$$F_n = A_n + B_n$$

Then for the first two members of the recurrent series F_n, A_n, and B_n should be

$$F_1 = 1 = A_1 + B_1$$

$$F_2 = 2 = A_2 + B_2$$

By exhaustive search, we get:

$$A_1 = 0, \quad A_2 = 1, \quad B_1 = 1, \quad B_2 = 1$$

Those A_n and B_n are also the Fibonacci sequences:

$$A_n = F_{n-2} \quad \text{and} \quad B_n = F_{n-1} \tag{V.10}$$

Returning to (V.8) and (V.9),

$$\beta = \lim_{n \to \infty} \frac{F_{n-2}}{F_n} = \frac{1}{\tau^2} \tag{V.11}$$

$$1 - \beta = \lim_{n \to \infty} \frac{F_{n-1}}{F_n} = \frac{1}{\tau}$$

Then the divergence coefficient will be $\beta = \dfrac{1}{\tau^2}$.

V.2.2. The phyllotactic pattern with divergence coefficient $1/\tau^2$

Unlike the previous problem, we know the divergence coefficient

$$\beta = \frac{1}{\tau^2} \quad \text{and} \quad 1-\beta = \frac{1}{\tau}, \tag{V.12}$$

but the number of spirals (parastichies) is unknown.

Let's choose an arbitrary node with the number M and located next to it, on the same spiral (parastichy), a node with the number K. From the fact that M and K are nearby nodes, it follows

$$|M - K| \to \min$$

The angle at which the radius vector rotates from the node M to the node K, gliding along the main genetic spiral, as follow from Figure V.1:

$$\Theta_m = 2\pi\beta |M - K|$$

And on the ancillary genetic spiral:

$$\Theta_a = 2\pi(1-\beta)|M - K|$$

The number of complete turns of the radius-vector when sliding along the main genetic spiral from the node M to the point M' (Figure V.1) is denoted by A, and by the ancillary genetic spiral is denoted by B. Moreover, A and B are integers. Now, the angle $|\Delta\theta|$ can be calculated in two ways:

$$|\Delta\theta| = |\Theta_m - 2\pi A| = 2\pi \left| \frac{|K-M|}{\tau^2} - A \right|$$

$$|\Delta\theta| = |2\pi B - \Theta_a| = 2\pi \left| B - \frac{|K-M|}{\tau} \right|$$

Because $|\Delta\theta| \to 0$, then

$$A \approx \frac{1}{\tau^2}|K - M| \tag{V.13}$$

$$B \approx \frac{1}{\tau}|K - M| \tag{V.14}$$

If we divide (V.13) by (V.14), we get:

$$B \approx \tau A \qquad (V.15)$$

Consequently:

$$A + B \approx A + \tau A = \tau^2 A \qquad (V.16)$$

From (V.15) and (V.16) it follows that for three integers A, B and $(A + B)$ will make the proportion:

$$(A+B):B:A \approx \tau^2 : \tau : 1 \qquad (V.17)$$

Consequently, A, B and $(A + B)$ are members of some generating recurrent sequence. From (V.12), (V.13), and (V.14), it follows that $|M - K| = (A + B) \rightarrow \min$, that is meaning, the sum of the initial members of this series should be minimal. As mentioned in Chapter II.2, the Fibonacci sequence has the minimum initial terms of all the generating recurrent sequences. Therefore:

$$\begin{cases} A = F_{n-2} \\ B = F_{n-1} \\ |K - M| = F_n \end{cases} \qquad (V.18)$$

From $|M - K| = F_n$ follows the difference in numbers between neighboring nodes on one spiral (parastichy) is equal to one of the Fibonacci numbers F_n, and number of spirals (parastichies) will also be F_n.

V.3. THE ANALYSIS OF THE MINIMUM DISTANCES BETWEEN THE NODES IN THE FIBONACCI PHYLLOTACTIC LATTICE

Let's find the angle $|\Delta\theta|$ at Figure V.1 for nearby nodes M and $M + F_n$. Let substitute (V.10) and (V.11) into (V.7):

$$|\Delta\theta| = \left| 2\pi \left(\frac{F_n}{\tau^2} - F_{n-2} \right) \right| \qquad (V.19)$$

Let's use Binet's formula (II.4) for (V.19):

$$|\Delta\theta| = \left| 2\pi \left(\frac{F_n}{\tau^2} - F_{n-2} \right) \right| = \left| 2\pi \left(\frac{\tau^{n+1} - (-1)^{n+1} \tau^{-(n+1)}}{\tau^2 \sqrt{5}} - \frac{\tau^{n-1} - (-1)^{n-1} \tau^{-(n-1)}}{\sqrt{5}} \right) \right| =$$

$$= \left| \frac{2\pi}{\sqrt{5}} (-(-1)^{n+1} \tau^{-n-3} + (-1)^{n-1} \tau^{-n+1}) \right| = \frac{2\pi}{\tau^{n+1}} \approx \frac{2\pi}{\sqrt{5}} \frac{1}{F_n} \qquad (V.20)$$

If $|\Delta\theta| \to 0$, then we can use the well-known approximation $\sin(l) \approx l$, where $l \to 0$ for (V.6):

$$\Delta T^2 = (M-K)^2 + MK(\Delta\theta)^2 \qquad (V.21)$$

Let's find ΔT for nearby nodes M and $M + F_n$:

$$\Delta T^2 \approx F_n^2 + M(M+F_n)\left(\frac{2\pi}{\tau^{n+1}}\right)^2 \approx F_n^2 + M(M+F_n)\left(\frac{2\pi}{\sqrt{5}}\frac{1}{F_n}\right)^2$$

For the large M, we can take $M \gg F_n$, then the approximate ΔT value can be calculated as:

$$\Delta T \approx \sqrt{F_n^2 + \left(\frac{2\pi}{\tau^{n+1}}\right)^2 M^2} \approx \sqrt{F_n^2 + \left(\frac{2\pi}{\sqrt{5}}\right)^2 \frac{M^2}{F_n^2}} \qquad (V.22)$$

From this formula, it follows that the distance between nearby nodes depends on the node number M and some parameter n, the meaning of which will be clarified below.

V.4. THE THREADS OF THE NODES OF THE PHYLLOTACTIC LATTICE AND THEIR VISUAL PERCEPTION

From (V.18), it follows that the node $M + F_n$ will be a nearby node to the node M. Also, the nodes $M + i \cdot F_n$ and $M + (i+1)F_n$ will be the nearby, where $i \in Z$.

A subset of the set of all nodes of the phyllotactic lattice with numbers is:

$$s + i \cdot F_n$$

where $i, n, s \in N$ and $0 \leq s < F_n$, will be called the thread of the nodes of the phyllotactic lattice or simply the thread.

The number n will be called the degree of the thread that owns the nodes of the phyllotactic lattice with numbers $s + i \cdot F_n$.

The combination of all threads with the same degree n will be called the n-th family of threads.

The properties of the threads of nodes of the phyllotactic lattice:

1. Each node of the phyllotactic lattice belongs to only one thread with degree n. Moreover, each node belongs to one thread from each n-th family of threads.
2. The number of different threads in the n-th family of threads is equal to F_n.
3. The set of nodes of the phyllotactic lattice belonging to one thread with degree n does not intersect with the set of nodes belonging to another thread with the same degree n.

4. The set of nodes belonging to all the threads of each n-th family of threads coincides with the set of all nodes of the phyllotactic lattice.
5. The angle of rotation of the radius-vector between the nearby nodes of the phyllotactic lattice, which is belonging to the same thread, is constant (V.20).

It is necessary here to dwell on the differences in the terms *the thread* and *parastichy*. These concepts are close, but not identical. *A thread*, just like a *node of a phyllotactic lattice*, is a mathematical abstraction, and parastichy is a combination of discrete botanical objects of the same type (primordia) that human perception unites into a spiral. For example, having seen the inflorescence of a sunflower, our perception unites sunflower seeds (primordia) in spirals, which are commonly called parastichies. As it was shown above, primordia belonging to one parastichy do not form an object with a specific entity.

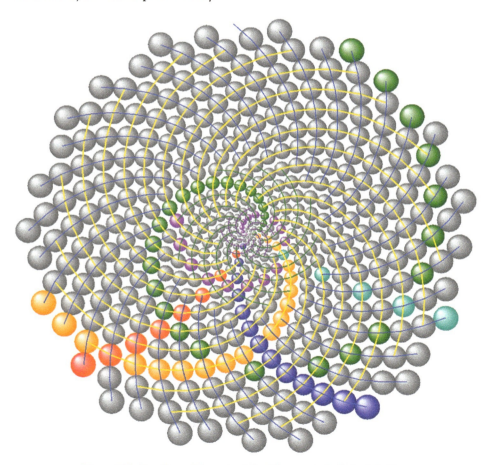

Figure V.2. Families of threads at the Fibonacci phyllotaxis lattice.

Figure V.2 shows the Fibonacci phyllotaxis lattice. Nodes belonging to the same thread are colored the same.

V.5. DIAPASONS AND NODES OF "THE PHYLLOTAXIS RISES"

As it was discussed above, the distance between nearby nodes depends on the node number M and the degree of the thread n. We construct a family of curves for each n according to (V.22). Each of these curves is a graph of the function of the distance between nearby nodes from the node number at a fixed n. To be more evident, we are going to assign each n to its own color. Then each n-th family of threads will have its own color curve. The resulting color curve family is depicted in Figure V.3.

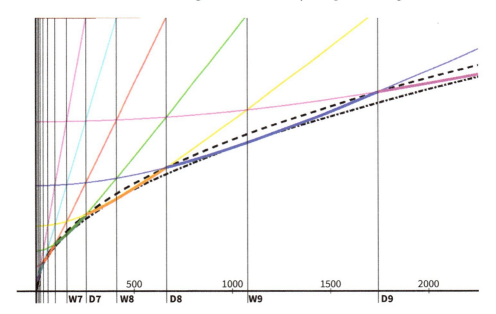

Figure V.3. The graph of the distance between nearby nodes for families of threads as the function of a node number.

Figure V.3 shows that the entire set of lattice nodes of the phyllotaxis is divided into subsets (diapasons) within which a curve of a certain color is closest to the ordinate axis. Let me explain with an example: for a node with the number 500 – the orange curve closest to the axis of ordinates – corresponds to a thread with the degree $n = 8$. And for a node with number 1000 – the blue curve closest to the axis of ordinates – corresponds to a thread with the degree $n = 9$. This means that for the node with the number 500, the nearby nodes will be the nodes with the numbers $500 + F_8$ and $500 − F_8$ (534 and 466); for the node with the number 1000, the nearby nodes are the nodes with the numbers $1000 + F_9$ and $1000 − F_9$ (1055 and 945).

The node corresponding to the intersection point of the graphs (Figure V.3) that correspond to the n-th and the $(n + 1)$-th families of threads will be called *the upper border of the n-th diapason* and denoted by D_n, and the node corresponding to the intersection point of the graphs that correspond to the $(n - 1)$-th and the n-th families of threads will be called *the lower border of the n-th diapason* and denoted by D_{n-1}. It is clear that the upper border of the $(n - 1)$-th diapason coincides with the lower border of the n-th diapason.

Accordingly, *the n-th diapason* will be called the set of nodes whose numbers are between the lower and upper borders of the n-th diapason.

As it follows from the definition, the node D_n belongs to the n-th and the $(n + 1)$-th diapasons. We can use (V.22) to calculate D_n:

$$F_{n+1}^2 + \left(\frac{2\pi}{\tau^{n+2}}\right)^2 D_n^2 = F_n^2 + \left(\frac{2\pi}{\tau^{n+1}}\right)^2 D_n^2 \tag{V.23}$$

$$(2\pi)^2 D_n^2 \left(\frac{1}{\tau^{2n+2}} - \frac{1}{\tau^{2n+4}}\right) = F_{n+1}^2 - F_n^2$$

$$D_n \approx \frac{\tau^{n+2} F_n}{2\pi} \approx \frac{\sqrt{5}\tau}{2\pi} F_n^2 \tag{V.24}$$

Table V.1. The borders of the diapasons and their corresponding families of threads.

The number of diapasons n	The number of threads in the n-th family F_n	Color of the n-th threads family according to Figure V.3	The lower border of the n-th diapason D_{n-1}	The upper border of the n-th diapason D_n
4	5	VIOLET	5	14
5	8	CYAN	14	37
6	13	RED	37	97
7	21	GREEN	97	254
8	34	ORANGE	254	665
9	55	BLUE	665	1742
10	89	MAGENTA	1742	4561
11	144	BROWN	4561	11941

Let us return to the visual perception of the phenomenon of phyllotaxis. Jean [3] distinguishes between "explicitly visible" and "implicitly visible" parastichies. The phrases "explicitly visible" and "implicitly visible" characterize the subjective perception of a human. However, it is now possible to formalize these terms, i.e. describe them mathematically. An "explicitly visible" parastichy will be a subset of the nodes of one of the threads of the n-th family, whose numbers are within the n-th diapason. In the future, we will color these nodes with a bold corresponding to the n-th family of threads (see Table V.1).

As observations show, human perception can simultaneously distinguish up to three different parastichies. As Figure V.3 shows, within the n-th diapason above the curve of the n-th family of threads at a short distance, there are the curves of the $(n-1)$-th and the $(n+1)$-th family of threads. To be exact, within the n-th diapason, the threads of the $(n-1)$-th and the $(n+1)$-th families will also be "seen", but not so clearly; this corresponds to the "implicitly visible" parastichies. The numerous nodes of the $(n-1)$-th and the $(n+1)$-th family of threads that are in the n-th diapason will be colored in the light color (see Table V.5.1).

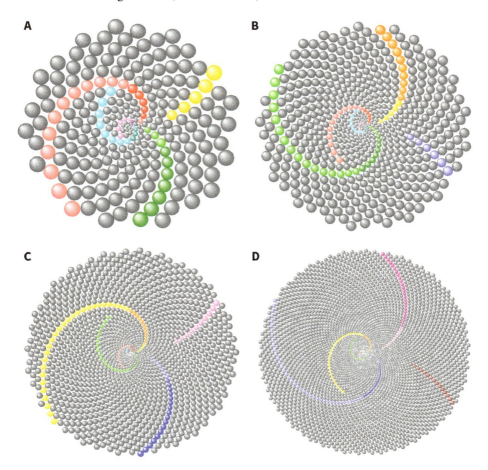

Figure V.4. The phyllotactic lattice at different scales with colored nodes:
A – 250 nodes, B – 600 nodes, C – 1600 nodes, D – 4000 nodes.

Figure V.4 shows that our perception distinguishes very well the "connection" between the nodes colored in the bold color and detects the "connection" between the nodes colored in the light. Consequently, the thread of the n-th family are clearly visible within the n-th diapason and are relatively visible within the $(n - 1)$-th and the $(n + 1)$-th diapasons. And, outside of the $(n - 1)$-th, the n-th, and the $(n + 1)$-th diapasons the thread of the n-th family are no longer visible.

Similarly to (V.23), we can find the node number W_n corresponding to the intersection point of the graphs corresponding to the $(n - 1)$-th and the $(n + 1)$-th families of threads:

$$F_{n+1}^2 + \left(\frac{2\pi}{\tau^{n+2}}\right)^2 W_n^2 = F_{n-1}^2 + \left(\frac{2\pi}{\tau^n}\right)^2 W_n^2 \qquad (V.25)$$

$$(2\pi)^2 W_n^2 \left(\frac{1}{\tau^{2n}} - \frac{1}{\tau^{2n+4}}\right) = F_{n+1}^2 - F_{n-1}^2$$

$$W_n \approx \frac{\tau^{n+1} F_n}{2\pi} \approx \frac{\sqrt{5}}{2\pi} F_n^2$$

As it is shown above, in the vicinity of nodes close to D_n, human perception clearly distinguishes two parastichies, in mathematical terms – the n-th and the $(n + 1)$-th family of threads. In the vicinity of the nodes close to W_n, human perception can distinguish three parastichies: the $(n - 1)$-th, the n-th, and the $(n + 1)$-th families of threads. At this point, there is a visual transition from the perception of two opposed parastichies with an index (F_{n-1}, F_n) to the perception of two opposed parastichies with an index (F_n, F_{n+1}). This visual effect is called "the phyllotaxis rises." Therefore, W_n will be called *the node of the phyllotaxis rises* (Table V.2). In reality, there is "rise" at all, the structure of the phyllotactic lattice remains unchanged.

Table V.2. Node of rise and the index of parastichies.

n	Node of rise W_n	Changing the index of parastichies $(F_{n-1}, F_n) \to (F_n, F_{n+1})$
5	23	$(5, 8) \to (8, 13)$
6	60	$(8, 13) \to (13, 21)$
7	157	$(13, 21) \to (21, 34)$
8	411	$(21, 34) \to (34, 55)$
9	1077	$(34, 55) \to (55, 89)$
10	2819	$(89, 144) \to (144, 233)$

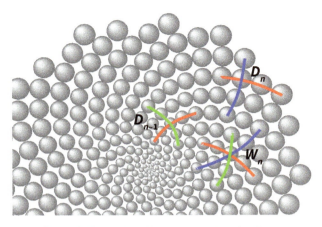

Figure V.5. Border of diapason and node of rise.

Figure V.5 explains the difference in the visual perception of the boarder of diapason D_n and the node of rise W_n. Near the D_n node are four nodes: $D_n - F_n$, $D_n + F_n$, $D_n - F_{n-1}$, $D_n + F_{n-1}$, forming a "cross". Near the node W_n. There will be six nodes: $W_n - F_n$, $W_n + F_n$, $W_n - F_{n+1}$, $W_n + F_{n+1}$, $W_n - F_{n-1}$, $W_n + F_{n-1}$, forming a hexagonal "snowflake."

V.6. DIAMETER OF AN ELEMENT OF THE PHYLLOTACTIC LATTICE

If we assume that primordia are similar to steel disks, then in terms of the DH-Model, each node of the phyllotactic lattice corresponds to one incompressible and non-deformable disk, the center of which coincides with this node, and the diameter is equal to the distance to the nearest node. Let us find the approximating function of the dependence radius of the disk on the node number. The graph of this function should pass through the points of the smallest distance between nodes in each diapason (dash-dotted line in Figure V.4). Figure V.4 shows these points have the ordinate of the node of rise W_n between the threads of the $(n-1)$-th and the $(n+1)$-th families (V.25) and the abscissa as the distance between the closest nodes for the n-th family in the n-th diapason (V.22):

$$\begin{cases} M = \dfrac{\sqrt{5}}{2\pi} F_n^2 \\ ds(M) = \sqrt{F_n^2 + \left(\dfrac{2\pi}{\sqrt{5}}\right)^2 \dfrac{M^2}{F_n^2}} \end{cases} \qquad (V.26)$$

Let's substitute the first equality $F_n^2 = \dfrac{2\pi}{\sqrt{5}} M$ into the second, to exclude n:

$$ds(M) = \sqrt{\dfrac{2\pi}{\sqrt{5}} M + \left(\dfrac{2\pi}{\sqrt{5}}\right)^2 M^2 \dfrac{\sqrt{5}}{2\pi M}} = \sqrt{\dfrac{2\pi}{\sqrt{5}} M + \dfrac{2\pi}{\sqrt{5}} M} = \sqrt{\dfrac{4\pi}{\sqrt{5}} M}$$

If we assume that primordia are similar to silicone disks, then in terms of the DH-Model, each node of the phyllotactic lattice corresponds to an incompressible, but deformable disk whose center coincides with this node, and these disks fill the space of the phyllotactic lattice to its maximum. The disk cannot be reduced in volume but can change its shape. For example, water cannot be compressed, but can change shape. Let us find the approximating function of the dependence of the radius of such a disk on the node number. The graph of this function passes through the border points of each diapason (V.24) (dashed line in Figure V.4):

$$\begin{cases} M = \dfrac{\sqrt{5}}{2\pi} \tau F_n^2 \\ d(M) = \sqrt{F_n^2 + \left(\dfrac{2\pi}{\sqrt{5}}\right)^2 \dfrac{M^2}{F_n^2}} \end{cases}$$

Let substitute the first equality $F_n^2 = \dfrac{2\pi}{\sqrt{5}} M$ into the second, to exclude n:

$$d(M) = \sqrt{\dfrac{2\pi}{\sqrt{5}\tau} M + \left(\dfrac{2\pi}{\sqrt{5}}\right)^2 M^2 \dfrac{\sqrt{5}\tau}{2\pi M}} = \sqrt{\dfrac{2\pi}{\sqrt{5}\tau} M + \dfrac{2\pi\tau}{\sqrt{5}} M} = \sqrt{2\pi M}$$

This formula was previously known as $d(M) = \sqrt{6M}$ and was used to draw a lattice of the planar Archimedean phyllotaxis.

As can be seen from numerous photographs of phyllotactic patterns, silicone discs are much closer to the real primordia. Therefore, we will consider the silicone disc only as a physical analogy of primordium. Therefore, the disk will be called *the element of the phyllotactic lattice*.

Let us find the packing density for L incompressible and deformable elements of the phyllotactic lattice as the ratio of the sum of the disk areas to the area of the entire the phyllotactic lattice:

$$\sum_{i=1}^{L} \pi \left(\dfrac{d_i}{2}\right)^2 : \pi L^2 = \dfrac{\pi}{2} \sum_{i=1}^{L} \left(\sqrt{i}\right)^2 : L^2 = \dfrac{\dfrac{\pi}{2} \dfrac{L(L-1)}{2}}{L^2} \approx \dfrac{\pi}{4} \quad (V.27)$$

V.7. THE ANGLE BETWEEN TANGENT LINES TO THE OPPOSED PARASTICHIES

One of the main ways of visual identification of opposed parastichy pair is the assumption that the tangent lines to parastichies intersect at a right angle [3]. Let's check this assumption for the DH-Model.

As discussed in Section V.4, the mathematical description of parastichy is the thread of lattice's nodes. Let us find the formula of the spiral passing through the nodes belonging to one thread from the n-th family.

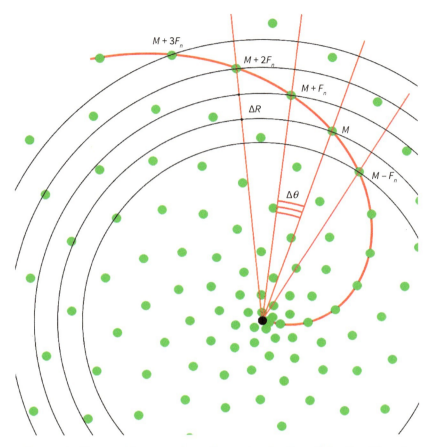

Figure V.6. The angel between the radius-vectors for the *n*-th family of threads.

Figure V.6 shows the rotation of the radius-vector, from the node M to the node $M + F_n$ by angle $\Delta\theta$ and the radius-vector length increases by ΔR. According to (V.20), $|\Delta\theta| = \dfrac{2\pi}{\tau^{n+1}}$ and is constant for the *n*-th family of threads. Also, according to (V.18), $\Delta R = F_n$ and is constant for the *n*-th family of threads. Therefore the nodes belonging to the thread of degree n will lie on one of the spirals:

$$\begin{cases} x(\theta) = \theta\cos\left(2\pi\left(\dfrac{(-1)^n}{F_n \tau^{n+1}}\theta + \dfrac{1}{F_n}s\right)\right) \\ y(\theta) = \theta\sin\left(2\pi\left(\dfrac{(-1)^n}{F_n \tau^{n+1}}\theta + \dfrac{1}{F_n}s\right)\right) \end{cases}, \quad (\text{V.28})$$

where $s \in N$ and $0 \leq s < F_n$.

Moreover, with $n = 1$ (V.28), it becomes the main genetic spiral and, with $n = 0$, it becomes ancillary genetic spiral.

Accordingly, the nodes belonging to the thread with $n + 1$ degree will lie on one of the spirals:

$$\begin{cases} x(\theta) = \theta\cos\left(2\pi\left(\dfrac{(-1)^{n+1}}{F_{n+1}\tau^{n+2}}\theta + \dfrac{1}{F_{n+1}}q\right)\right) \\ y(\theta) = \theta\sin\left(2\pi\left(\dfrac{(-1)^{n+1}}{F_{n+1}\tau^{n+2}}\theta + \dfrac{1}{F_{n+1}}q\right)\right) \end{cases}, \quad \text{(V.29)}$$

where $q \in N$ and $0 \le q < F_{n+1}$.

Figure V.7 shows the nodes of the Archimedean phyllotactic lattice (green circles and two spirals red and blue). On the red spiral, there are the nodes belonging to the thread with degree $n = 6$; on the blue one, there are the nodes belonging to the thread with degree $n = 7$. The radius-vector (the black line) passes through the point of intersection of these spirals. Tangent lines are drawn on both spirals through the same point. As can be seen on Figure V.7, the angle between the tangent lines is equal to the sum of the angles between the radius-vector and these tangent lines.

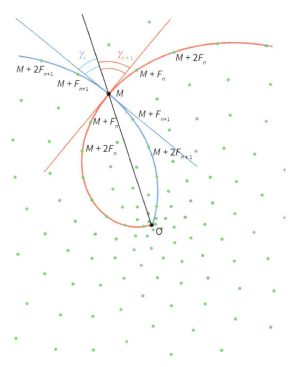

Figure V.7. The angle between tangent lines to the *n*-th and (*n* + 1)-th thread's spirals.

The Planar DH-Model on Archimedean Spirals 43

We know the remarkable property of an Archimedean spiral from analytic geometry (Appendix D, Formula D.6) that the tangent function of the angle between the tangent line and the radius vector passing through the tangency point is equal to the angle of rotation of the entire spiral. From (V.28) and (V.29) follows:

$$\tan(\gamma_n) = \left| 2\pi \frac{(-1)^n}{F_n \tau^{n+1}} \theta \right|, \qquad \tan(\gamma_{n+1}) = \left| 2\pi \frac{(-1)^{n+1}}{F_{n+1} \tau^{n+2}} \theta \right|$$

Consequently:

$$\gamma(M) = \gamma_n + \gamma_{n+1} = arctan\left(\frac{2\pi M}{F_n \tau^{n+1}} \right) + arctan\left(\frac{2\pi M}{F_{n+1} \tau^{n+2}} \right)$$

Now we can find the angle between tangent lines to the threads of the n-th and the $(n+1)$-th families at the node D_n, which is the upper border of the n-th diapason, using (V.24)

$$\gamma(D_n) = arctan(\tau) + arctan\left(\frac{1}{\tau} \right) = \frac{\pi}{2}$$

This fully confirms the earlier empirically obtained statement that tangent lines to opposed parastichies intersect at a right angle [3].

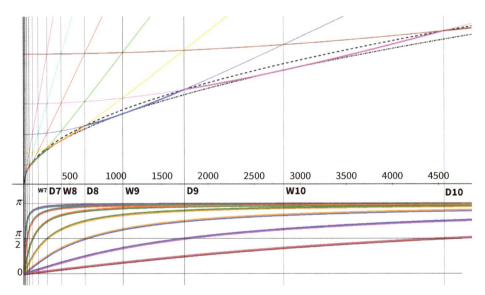

Figure V.8. The combined graphs of the distance between nearby nodes for families of threads and the angle between tangent lines to thread's spirals as the functions of a node number.

Figure V.8 displays the combined graphs of the distance between the adjacent nodes of the threads and the angle between tangent lines. Very close to the border of the *n*-th range, the angle between the tangent lines is close to a right angle and there we can evidently see the pair of oppositely twisted parastichies with the index (F_n, F_{n+1}). Perhaps that is why Jean [3] called them a "visible opposed parastichy pair."

V.8. IS PHYLLOTAXIS A FRACTAL OR A RECURSIVE STRUCTURE?

There is no strict definition of the term "a fractal" [16]. Therefore, we check the structure of the phyllotactic lattice for compliance with the intuitive notion of *self-similarity*. By self-similarity of the structure of the phyllotactic lattice, we will understand the existence of an affine transformation, which changes every *M*-th element of the phyllotactic lattice into (*M* + 1)-th, i.e. the phyllotactic lattice is transformed into itself. However, the planar Archimedean spiral cannot be transformed into itself; therefore, this Archimedean phyllotactic lattice is not self-similarity.

By the recursiveness of the structure, we will mean here the existence of some algorithm that transforms the *M*-th element of the Archimedean phyllotactic lattice into (*M* + 1)-th.

Let's consider the *M*-th element in the lattice structure of Archimedean phyllotaxis, which has coordinates and diameter:

$$\begin{cases} x(M) = M\cos(2\pi\beta M) \\ y(M) = M\sin(2\pi\beta M) \\ d(M) = \sqrt{2\pi M} \end{cases}$$

We can express M through $x(M)$ and $y(M)$:

$$M = \sqrt{x^2(M) + y^2(M)} \quad \text{and} \quad tg(2\pi\beta M) = \frac{y(M)}{x(M)}$$

Therefore, the coordinates and diameter of the (*M* + 1)-th element will be

$$\begin{cases} x(M+1) = \left(1 + \sqrt{x^2(M) + y^2(M)}\right)\cos\left(2\pi\beta + arctg\left(\frac{y(M)}{x(M)}\right)\right) \\ y(M+1) = \left(1 + \sqrt{x^2(M) + y^2(M)}\right)\sin\left(2\pi\beta + arctg\left(\frac{y(M)}{x(M)}\right)\right) \\ d(M+1) = \sqrt{2\pi\left(1 + \sqrt{x^2(M) + y^2(M)}\right)} \end{cases}$$

That means that the $(M + 1)$-th element can always be calculated from M-th and the algorithm will be the same for all elements of the Archimedean phyllotactic lattice.

V.9. THE PLANAR DH-MODEL FOR "ACCESSORY" SEQUENCES

Jean [3] gives the formula $\dfrac{2\pi}{t + \tau^{-1}}$ for calculating the divergence angle for "accessory" sequences:

$$1, t, t+1, t+1, 3t+2, 5t+3, \ldots, \text{ were } t > 1, t \in N \qquad (V.30)$$

Let us verify this statement by finding the formula for calculating the divergence coefficient for Archimedean phyllotactic lattices generated by the Fibonacci-like sequences:

$$1, C_2, C_2+1, 2C_2+1, 3C_2+2, 5C_2+3, 8C_2+5, \ldots, F_{n-2}C_2 + F_{n-3}, \ldots$$

As follows from the Bravais-Bravais theorem, the nodes M and $M + C_n$ will be nearby and belong to the same thread. Similarly to Section V.2.1, we will find the angles through which the radius-vector rotates from the node M to the node $M + C_n$, sliding along the main and the ancillary genetic spirals:

$$\Theta_m = 2\pi\beta\left((M+C_n)-M\right) = 2\pi\beta C_n$$

$$\Theta_a = 2\pi(1-\beta)\left((M+C_n)-M\right) = 2\pi(1-\beta)C_n$$

The number of complete turns of the radius-vector from the node M to the point M' (Figure V.1) along the main genetic spiral is denoted as A_n, and by the ancillary genetic spiral B_n. The numbers A_n and B_n are integers because they figure out the number of complete rotations around the center. Therefore, the angle $|\Delta\theta|$ can be calculated in two ways:

$$|\Delta\theta| = |\Theta_m - 2\pi A_n| = 2\pi|\beta C_n - A_n| \qquad (V.31)$$

$$|\Delta\theta| = |2\pi B_n - \Theta_a| = 2\pi|B_n - (1-\beta)C_n|$$

Because $|\Delta\theta| \to 0$, then:

$$A_n \approx \beta C_n \qquad (V.32)$$

$$B_n \approx (1-\beta)C_n \qquad (V.33)$$

It follows that A_n and B_n are members of some recurrent series. Moreover, $A_n < B_n$ because $0 < \beta < 0.5$. If we sum (V.32) and (V.33), we get

$$C_n = A_n + B_n$$

Then for the first two members of the "accessory" sequence (V.30) should be performed:

$$C_1 = 1 = A_1 + B_1$$
$$C_2 = t = A_2 + B_2$$

Those A_n and B_n are integer recurrent sequences, by exhaustive search we find:

$$A_1 = 0, \quad A_2 = 1, \quad B_1 = 1, \quad B_2 = t - 1$$

Consequently:

$$A_n = F_{n-2}$$

From (V.32) and (II.9), we can find a formula for calculating the divergence coefficient for the phyllotactic lattice generated by the Fibonacci-like sequence:

$$\beta = \lim_{n \to \infty} \frac{A_n}{C_n} = \lim_{n \to \infty} \frac{F_{n-2}}{C_n} = \frac{1}{C_2 + \tau^{-1}} = \frac{1}{t + \tau^{-1}},$$

which completely coincides with the divergence angle formula for the "accessory" sequence from [3].

Let's use Binet's formulas (II.4) and (II.7) to calculate $|\Delta\theta|$ from (V.31):

$$|\Delta\theta| = 2\pi \left| \frac{C_n}{C_2 + \tau^{-1}} - F_{n-2} \right| = \frac{2\pi}{\sqrt{5}} \left| \tau^{n-1} - \frac{(-1)^{n-1} \tau^{-n+1}(C_2 - \tau)}{C_2 + \tau^{-1}} - \left(\tau^{n-1} - (-1)^{n-1} \tau^{-n+1} \right) \right| =$$

$$= \frac{2\pi}{\sqrt{5}} \tau^{-n+1} \left(\frac{C_2 + \frac{1}{\tau} - C_2 + \tau}{C_2 + \tau^{-1}} \right) = \frac{2\pi}{\sqrt{5}} \tau^{-n+1} \left(\frac{\sqrt{5}}{C_2 + \tau^{-1}} \right) \approx \frac{2\pi}{\sqrt{5}} \frac{1}{C_n} \quad \text{(V.34)}$$

Substitute (V.34) into (V.21):

$$\Delta T^2 = C_n^2 + M(M - C_n) \left(\frac{\pi}{\sqrt{5}} \frac{1}{C_n} \right)^2$$

For large M, we can take $M \gg C_n$ then the approximate ΔT value can be calculated as:

$$\Delta T \approx \sqrt{C_n^2 + \left(\frac{2\pi}{\sqrt{5}}\right)^2 \frac{M^2}{C_n^2}} \qquad (V.35)$$

In the same way as it is in Fibonacci phyllotaxis, nearby nodes of the phyllotactic lattice, generated the by the Fibonacci-like sequence, form threads:

$$s + i \cdot C_n$$

where $i, n, s \in N$ and $0 \leq s < C_n$.

The nodes belonging to the thread of degree n will lie on one of the spirals:

$$\begin{cases} x(\theta) = \theta \cos\left(2\pi\left(\frac{(-1)^n}{C_n \tau^{n-1}} \beta\theta + \frac{1}{C_n} s\right)\right) \\ y(\theta) = \theta \sin\left(2\pi\left(\frac{(-1)^n}{C_n \tau^{n-1}} \beta\theta + \frac{1}{C_n} s\right)\right) \end{cases} \qquad (V.36)$$

It should be noted that here, the thread with degree $n = 1$ (V.28) also becomes the main genetic spiral, and the thread with degree with $n = 0$ becomes ancillary genetic spiral.

Just as in Section V.5, we will find the borders of the n-th diapason:

$$C_{n+1}^2 + \left(\frac{2\pi}{\sqrt{5}}\right)^2 \frac{1}{C_{n+1}^2} D_n^2 = C_n^2 + \left(\frac{2\pi}{\sqrt{5}}\right)^2 \frac{1}{C_n^2} D_n^2$$

$$D_n = \frac{\sqrt{5}}{2\pi} C_{n+1} C_n \approx \frac{\tau^n}{2\pi\beta} C_n \qquad (V.37)$$

and nodes of rise:

$$C_{n-1}^2 + \left(\frac{2\pi}{\sqrt{5}}\right)^2 \frac{W_n^2}{C_{n-1}^2} = C_{n+1}^2 + \left(\frac{2\pi}{\sqrt{5}}\right)^2 \frac{W_n^2}{C_{n+1}^2}$$

$$W_n = \frac{\sqrt{5}}{2\pi} C_{n-1} C_{n+1} \approx \frac{\sqrt{5}}{2\pi} C_n^2 \qquad (V.38)$$

Like in Section V.6, we will find the function of dependence of the radius of an incompressible, but deformable element of the phyllotactic lattice on the node number:

$$\begin{cases} M = \dfrac{\sqrt{5}}{2\pi}\tau C_n^2 \\ d(M) = \sqrt{C_n^2 + \left(\dfrac{2\pi}{\sqrt{5}}\right)^2 \dfrac{M^2}{C_n^2}} \end{cases}$$

We will substitute the first equality $C_n^2 = \dfrac{2\pi}{\sqrt{5}\tau}M$ into the second to exclude n:

$$d(M) = \sqrt{\dfrac{2\pi}{\sqrt{5}\tau}M + \left(\dfrac{2\pi}{\sqrt{5}}\right)^2 M^2 \dfrac{\sqrt{5}\tau}{2\pi M}} = \sqrt{\dfrac{2\pi}{\sqrt{5}\tau}M + \dfrac{2\pi\tau}{\sqrt{5}}M} = \sqrt{2\pi M}$$

Find the packing density for L incompressible and deformable elements, similarly to (V.27):

$$\sum_{i=1}^{L} \pi\left(\dfrac{d_i}{2}\right)^2 : \pi L^2 = \dfrac{\pi}{2}\sum_{i=1}^{L}\left(\sqrt{i}\right)^2 : L^2 = \dfrac{\pi \dfrac{L(L-1)}{2}}{L^2} \approx \dfrac{\pi}{4}$$

Like in Section V.7, we find the angle between tangent lines to the opposed parastichies by using the formula of spirals (V.36) on which the threads of the nodes of the phyllotactic lattice lie, generated by the Fibonacci-like sequences.

$$\gamma(M) = \arctan\left(\dfrac{2\pi}{C_n\tau^{n-1}}\beta M\right) + \arctan\left(\dfrac{2\pi}{C_{n+1}\tau^n}\beta M\right) \qquad (V.39)$$

If in (V.39) we substitute the border of the diapason from (V.37), then the angle between tangent lines to the opposed parastichy pair will be a right angle:

$$\gamma(D_n) = \arctan(\tau) + \arctan\left(\dfrac{1}{\tau}\right) = \dfrac{\pi}{2}$$

If we substitute $C_2 = 2$ into formulas (V.34)–(V.39), then they will turn into the corresponding formulas for the phyllotactic lattice generated by the Fibonacci sequence. This proves that the phyllotactic lattice generated by the Fibonacci sequence is a particular case of the phyllotactic lattice generated by the Fibonacci-like sequence.

The chart below shows us phyllotactic lattices generated by the Fibonacci-like sequences with different C_2.

C_2	Figure	URL	C_2	Figure	URL
2	Figure V.9	https://youtu.be/gW7Mg1OPXT0	12	Figure V.19	https://youtu.be/Or_BroIXuvc
3	Figure V.10	https://youtu.be/9DH2UID6WXk	13	Figure V.20	https://youtu.be/BC_a0qL1FT4
4	Figure V.11	https://youtu.be/FfAZSh2Vxek	14	Figure V.21	https://youtu.be/qC2kF8Y5Sp4
5	Figure V.12	https://youtu.be/AQfV4ZWOUeI	15	Figure V.22	https://youtu.be/iJE6ZO6sWYI
6	Figure V.13	https://youtu.be/t3JYKButl_M	16	Figure V.23	https://youtu.be/RqXOPdc0eXY
7	Figure V.14	https://youtu.be/6Ddwgz0-t8g	17	Figure V.24	https://youtu.be/cqxPQU3q39E
8	Figure V.15	https://youtu.be/7VguB-5s9FM	18	Figure V.25	https://youtu.be/LptGa5Tmk0E
9	Figure V.16	https://youtu.be/BX-TO8SpPjw	19	Figure V.26	https://youtu.be/uYwteXHrNAY
10	Figure V.17	https://youtu.be/8ninScMKOA0	20	Figure V.27	https://youtu.be/f45vX9njm8M
11	Figure V.18	https://youtu.be/HGJoN9wRXK8			

V.10. THE PLANAR DH-MODEL FOR "LATERAL" SEQUENCES

Jean [3] defines the next recurrent sequences as "lateral" sequences:

$$2, 2t+1, 2t+3, 4t+4, 6t+7, \ldots, \text{ where } t \geq 2, t \in \mathbb{N} \qquad (V.40)$$

Phyllotactic patterns generated by these sequences are called "anomalous" phyllotaxis. The formula for finding the divergence angle for these patterns is also given:

$$b = 2\pi \left(2 + \frac{1}{t+\tau^{-1}}\right)^{-1} \quad \text{or} \quad b = 2\pi \frac{t+\tau^{-1}}{(2t+1)+2\tau^{-1}} \qquad (V.41)$$

Let's look at phyllotactic lattices generated by non-multiple recurrent sequences:

$$G_1, G_2, G_2+G_1, 2G_2+G_1, 3G_2+2G_1, 5G_2+3G_1, \ldots,$$
where $1 < G_1, 2G_1 \leq G_2$; $\gcd(G_1, G_2) = 1$, $G_1, G_2 \in \mathbb{N}$.

These sequences are called non-multiple because their initial terms G_1 and G_2 do not have a common multiple divisor. The recurrent sequence (V.40) with the initial terms $G_1 = 2$ and $G_2 = 2t+1$ is a particular case of non-multiple recurrent sequences.

Let us find the formula for the coefficient of divergence for the phyllotactic lattice generated by a non-multiple recurrent sequence. If in the phyllotactic pattern we see G_n parastichy of one family, then, according to Bravais-Bravais theorem, the nodes M and $M + G_n$ will be nearby and belong to the same thread.

Similarly to Section V.2.1, the number of complete turns of the radius-vector from the node M to the point M' (Figure V.1) along the main genetic spiral is denoted as A_n, and by the ancillary genetic spiral B_n. The numbers A_n and B_n are integers because they figure out number of complete rotations around the center. Therefore the angle $|\Delta\theta|$ can be calculated in two ways:

$$|\Delta\theta| = |\Theta_m - 2\pi A_n| = 2\pi|\beta G_n - A_n| \tag{V.42}$$

$$|\Delta\theta| = |2\pi B_n - \Theta_a| = 2\pi|B_n - (1-\beta)G_n|$$

Because $|\Delta\theta| \to 0$, then:

$$A_n \approx \beta G_n \tag{V.43}$$

$$B_n \approx (1-\beta)G_n \tag{V.44}$$

It follows that A_n and B_n are members of some integral recurrent series. Moreover, $A_n < B_n$ because $0 < \beta < 0.5$. If we sum (V.43) and (V.44) we get

$$G_n = A_n + B_n$$

It follows from the last formula that A_n and B_n are some recurrent sequences, and their sum is equal to the generating non-multiple recurrent sequence G_n. Then, by analogy with (II.10)–(II.14), the divergence coefficient will be:

$$\beta = \lim_{n \to \infty} \frac{A_n}{G_n} = \frac{A_2 + \dfrac{A_1}{\tau}}{G_2 + \dfrac{G_1}{\tau}} \tag{V.45}$$

Let's use Binet's formulas (II.4) and (II.7) to calculate $|\Delta\theta|$ from (V.40):

$$|\Delta\theta_n| = 2\pi \left| \left(\frac{A_2 + A_1 \cdot \tau^{-1}}{\sqrt{5}} \tau^{n-1} - \frac{A_2 - A_1 \cdot \tau}{\sqrt{5}}(-\tau)^{-n+1} \right) \right.$$

$$\left. - \frac{A_2 + \dfrac{A_1}{\tau}}{G_2 + \dfrac{G_1}{\tau}} \left(\frac{G_2 + G_1 \cdot \tau^{-1}}{\sqrt{5}} \tau^{n-1} - \frac{G_2 - G_1 \cdot \tau}{\sqrt{5}}(-\tau)^{-n+1} \right) \right| =$$

$$= \frac{2\pi}{\sqrt{5}} \frac{\tau^{-n+1}}{G_2 + \frac{G_1}{\tau}} \left| \left(G_2 + \frac{G_1}{\tau} \right) (A_2 - A_1 \cdot \tau) - \left(A_2 + \frac{A_1}{\tau} \right) (G_2 - G_1 \cdot \tau) \right| \approx \frac{2\pi}{\sqrt{5} G_n} |A_1 G_2 - A_2 G_1|$$

(V.46)

$\Delta\theta_n$ cannot be equal to 0, because otherwise, the phyllotactic lattice disappears completely. Therefore:

$$|A_1 G_2 - A_2 G_1| \neq 0$$

Since A_1, A_2, G_1 and G_2 are integers, then $|\Delta\theta_n|$ will be minimal when

$$|A_1 G_2 - A_2 G_1| = 1 \qquad (V.47)$$

In fact, the equation (V.44) are two Linear Diophantine equations:

$$\begin{cases} A_1 G_2 - A_2 G_1 = 1 \\ A_1 G_2 - A_2 G_1 = -1 \end{cases} \qquad (V.48)$$

Each of these equations has one and only one solution, which can be found using the Euclidean algorithm [17]. An example of using the Euclidean algorithm for solving Linear Diophantine equations is shown in Appendix E.

Let's substitute $A_1 = G_1 - B_1$ and $A_2 = G_2 - B_2$ in (V.48), then we get:

$$(G_1 - B_1) G_2 - (G_2 - B_2) G_1 = -1$$
$$B_1 G_2 - B_2 G_1 = 1 \qquad (V.49)$$

When comparing (V.48) to (V.49), we can come to the conclusion that A_n and B_n are mutually interchangeable and the two solutions of equation (V.44) correspond to two genetic spirals — the main and the ancillary.

The divergence coefficient for the phyllotactic lattice generated by non-multiple recurrent sequences can be calculated from the system:

$$\begin{cases} |A_1 G_2 - A_2 G_1| = 1 \\ \beta = \dfrac{A_2 + \dfrac{A_1}{\tau}}{G_2 + \dfrac{G_1}{\tau}} \end{cases}$$

Let's check the correctness of this algorithm for "lateral" sequences (V.40), in which $G_1 = 2$ and $G_2 = 2t + 1$, and let's find A_1 and A_2:

1. $A_1(2t+1) - 2A_2 = 1$, by the Euclidean algorithm $A_1 = 1$, $A_2 = t$,
2. $A_1(2t+1) - 2A_2 = -1$, by the Euclidean algorithm $A_1 = 1$, $A_2 = t + 1$.

The first option is smaller. Substitute $G_1 = 2$, $G_2 = 2t + 1$, $A_1 = 1$, $A_2 = t$ in (V.45):

$$\beta = \lim_{n \to \infty} \frac{A_n}{G_n} = \frac{A_2 + \dfrac{A_1}{\tau}}{G_2 + \dfrac{G_1}{\tau}} = \frac{t + \dfrac{1}{\tau}}{2t + 1 + \dfrac{2}{\tau}}$$

The resulting formula for the divergence coefficient completely coincides with (V.41).

Appendix C provides a table of the dependence of the divergence coefficient β, and A_1, A_2, B_1, B_2 from G_1 and G_2 for non-multiple recurrent sequences.

Returning to (V.46): if $|A_1 G_2 - A_2 G_1| = 1$, then

$$|\Delta\theta_n| \approx \frac{2\pi}{\sqrt{5} G_n} \tag{V.50}$$

For large M, we can take $M \gg G_n$ then the approximate ΔT value can be calculated as:

$$\Delta T \approx \sqrt{G_n^2 + \left(\frac{2\pi}{\sqrt{5 G_n^2}}\right)^2 M^2} \tag{V.51}$$

In the same way as it is in Fibonacci phyllotaxis, nearby nodes of the phyllotactic lattice, generated by non-multiple recurrent sequences, form threads:

$$s + i \cdot G_n$$

where $i, n, s \in \mathbb{N}$ and $0 \leq s < G_n$.

The nodes belonging to the thread of degree n will lie on one of these spirals:

$$\begin{cases} x(\theta) = \theta \cos\left(2\pi\left(\dfrac{(-1)^n}{G_n \tau^{n-1}} \beta\theta + \dfrac{1}{G_n} s\right)\right) \\ y(\theta) = \theta \sin\left(2\pi\left(\dfrac{(-1)^n}{G_n \tau^{n-1}} \beta\theta + \dfrac{1}{G_n} s\right)\right) \end{cases} \tag{V.52}$$

It should be noted here that the thread with degree $n = 1$ (V.28) also becomes the main genetic spiral, and the thread with degree with $n = 0$ also becomes the ancillary genetic spiral

Same as in Section V.5, we will find the borders of the n-th diapason.

$$G_{n+1}^2 + \left(\frac{2\pi}{\sqrt{5}}\right)^2 \frac{1}{G_{n+1}^2} D_n^2 = G_n^2 + \left(\frac{2\pi}{\sqrt{5}}\right)^2 \frac{1}{G_n^2} D_n^2$$

$$D_n = \frac{\sqrt{5}}{2\pi} G_{n+1} G_n \approx \frac{\tau^n}{2\pi\beta} G_n \qquad (V.53)$$

and nodes of rise

$$G_{n-1}^2 + \left(\frac{2\pi}{\sqrt{5}}\right)^2 \frac{W_n^2}{G_{n-1}^2} = G_{n+1}^2 + \left(\frac{2\pi}{\sqrt{5}}\right)^2 \frac{W_n^2}{G_{n+1}^2}$$

$$W_n = \frac{\sqrt{5}}{2\pi} G_{n-1} G_{n+1} \approx \frac{\sqrt{5}}{2\pi} G_n^2 \qquad (V.54)$$

Same as in Section V.6, we will find the function of dependence of the radius of incompressible and deformable element of the phyllotactic lattice on the node number:

$$\begin{cases} M = \dfrac{\sqrt{5}}{2\pi}\tau G_n^2 \\ d(M) = \sqrt{G_n^2 + \left(\dfrac{2\pi}{\sqrt{5}}\right)^2 \dfrac{M^2}{G_n^2}} \end{cases}$$

We substitute the first equality $G_n^2 = \dfrac{2\pi}{\sqrt{5}\tau} M$ into the second to exclude n:

$$d(M) = \sqrt{\frac{2\pi}{\sqrt{5}\tau} M + \left(\frac{2\pi}{\sqrt{5}}\right)^2 M^2 \frac{\sqrt{5}\tau}{2\pi M}} = \sqrt{\frac{2\pi}{\sqrt{5}\tau} M + \frac{2\pi\tau}{\sqrt{5}} M} = \sqrt{2\pi M}$$

Let's find the packing density for L incompressible and deformable elements, similarly to (V.27):

$$\sum_{i=1}^{L} \pi \left(\frac{d_i}{2}\right)^2 : \pi L^2 = \frac{\pi}{2}\sum_{i=1}^{L}\left(\sqrt{i}\right)^2 : L^2 = \frac{\pi \dfrac{L(L-1)}{2}}{L^2} \approx \frac{\pi}{4}$$

Same as in Section V.7, we find the angle between tangent lines to the opposed parastichies, using the formula of spirals (V.52) on which the threads of the nodes of the phyllotactic lattice lie, generated by non-multiple recurrent sequences.

$$\gamma(M) = \arctan\left(\frac{2\pi}{G_n \tau^{n-1}}\beta M\right) + \arctan\left(\frac{2\pi}{G_{n+1}\tau^n}\beta M\right) \qquad (V.55)$$

If in (V.55) we substitute the border of the diapason from (V.53), then the angle between tangent lines to the opposed parastichy pair will be a right angle:

$$\gamma(D_n) = arctan(\tau) + arctan\left(\frac{1}{\tau}\right) = \frac{\pi}{2}$$

If $G_1 = 1$, then from (V.44) follows $A_1 = 0$ and $A_2 = 1$.

If we substitute $G_1 = 1$ into formulas (V.45)–(V.55), then they will turn into corresponding formulas for the phyllotactic lattice generated by the Fibonacci-like sequence. This proves that the phyllotactic lattice generated by the Fibonacci-like sequence is a particular case of the phyllotactic lattice generated by the non-multiple recurrent sequences.

In the figures below, you can see the summary chart for phyllotactic lattices generated by Fibonacci-like sequences with different initial terms.

G_1	G_2	Figure	URL
2	5	Figure V.28	https://youtu.be/R4U3WlWMOQM
2	7	Figure V.29	https://youtu.be/vAX7TcaCRTE
2	9	Figure V.30	https://youtu.be/NHaD1JmRttY
2	11	Figure V.31	https://youtu.be/B-j7FqLd11k
2	13	Figure V.32	https://youtu.be/COBmv0tYrDI
2	15	Figure V.33	https://youtu.be/qpH7Bs_wx2E
2	17	Figure V.34	https://youtu.be/wvVU63TMwBs
2	19	Figure V.35	https://youtu.be/FcO_eipPq2I
2	21	Figure V.36	https://youtu.be/5E5M_6WbcLk
3	7	Figure V.37	https://youtu.be/qAiOQgqtpNs
3	8	Figure V.38	https://youtu.be/8k7V0i7MDuc
3	10	Figure V.39	https://youtu.be/zJBMEKMgsYQ
3	11	Figure V.40	https://youtu.be/bpPRxj1cO4c
3	13	Figure V.41	https://youtu.be/XBsvea_jzkU
3	14	Figure V.42	https://youtu.be/RuQhWXpvNCU
3	16	Figure V.43	https://youtu.be/IC1MtBTIHgg
3	17	Figure V.44	https://youtu.be/rhUOnotIE54
3	19	Figure V.45	https://youtu.be/veTahIJyucM
4	9	Figure V.46	https://youtu.be/HkNEy_ncV00

G_1	G_2	Figure	URL
4	11	Figure V.47	https://youtu.be/-6TE-PrbaDM
4	13	Figure V.48	https://youtu.be/jcgMy-dvbjc
4	15	Figure V.49	https://youtu.be/Y8ZZF21m9ow
4	17	Figure V.50	https://youtu.be/Ue3kkaJthzk
5	11	Figure V.51	https://youtu.be/4VLwNuJeRgM
5	12	Figure V.52	https://youtu.be/U3SW4neNKjY
5	13	Figure V.53	https://youtu.be/hxlySGn5EjQ
5	14	Figure V.54	https://youtu.be/cUdQSnq-NKU
5	16	Figure V.55	https://youtu.be/xmH7VaNr9Co
5	17	Figure V.56	https://youtu.be/ThZ3RlwRlto
5	18	Figure V.57	https://youtu.be/SLvlA8YXvog
5	19	Figure V.58	https://youtu.be/DXdgl00frik
5	21	Figure V.59	https://youtu.be/3KphO11Mrd4
5	22	Figure V.60	https://youtu.be/FRTzhvh21wA
5	23	Figure V.61	https://youtu.be/TX8vbksZhN4
5	24	Figure V.62	https://youtu.be/687_KeymvuM
6	13	Figure V.63	https://youtu.be/4mWT-AfD7Ms
6	17	Figure V.64	https://youtu.be/gf6TlQMEt10
6	19	Figure V.65	https://youtu.be/ov-yMu146Eg
7	15	Figure V.66	https://youtu.be/jDEyRXf7Sjk
7	16	Figure V.67	https://youtu.be/XX5M3sSDx18
7	17	Figure V.68	https://youtu.be/2YWT7lZULUM
7	18	Figure V.69	https://youtu.be/JFpzXog8CN4
7	19	Figure V.70	https://youtu.be/r_5VFtwRjeM
7	20	Figure V.71	https://youtu.be/0BsSHpQ5BHM
7	22	Figure V.72	https://youtu.be/6j1FVHpAdDY
8	17	Figure V.73	https://youtu.be/klyWsrhtpAs

The phyllotactic lattices above (Figure V.28 — Figure V.73) may seem to be the result of a mathematical fantasy, which has no analogies in nature. However, the similarity of the phyllotactic lattice generated by the recurrent sequence with initial terms 5 and 23 in Figure V.74 and the Spiral Aloe pattern in Figure V.75 suggests that there may be other phyllotactic patterns that are not mentioned in [3].

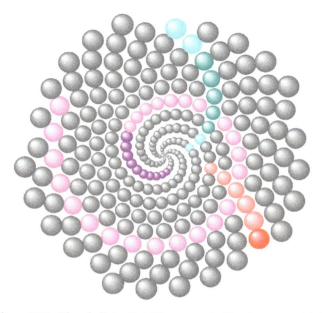

Figure V.74. The phyllotactic lattice generated by the non-multiple recurrent sequence $G_1 = 5$, $G_2 = 23$.

Figure V.75. Spiral Aloe.

V.11. CONVERSE OF THE FUNDAMENTAL THEOREM OF PHYLLOTAXIS

Let's consider the statement which is opposite to the fundamental theorem of phyllotaxis in terms of the DH-Model:

Let the phyllotactic pattern be generated by the allowed non-multiple recurrent sequence ($1 \leq G_1$, $2G_1 \leq G_2$, $\gcd(G_2, G_1) = 1$), and this pattern has a visible opposed parastichy pair with index (G_n, G_{n+1}). Therefore the coefficient of discrepancy β will lie in the interval $\left[\dfrac{u}{G_{n+1}}, \dfrac{v}{G_n}\right]$ ($v, u \in N$, $0 \leq v < G_n$, $0 \leq u < G_{n+1}$), where the encyclic numbers v and u can be found by solving the integer equation:

$$|vG_{n+1} - uG_n| = 1 \tag{V.56}$$

From the last formula, it follows that each parastichy pair (G_n, G_{n+1}) corresponds to a pair of encyclic numbers v and u. Let's denote a pair of encyclic numbers that correspond to the pair of numbers G_n and G_{n+1}, as u_n and v_n; then (V.56) will look like:

$$|v_n G_{n+1} - u_n G_n| = 1 \tag{V.57}$$

And for an opposed parastichy pair (G_{n+1}, G_{n+2}), the formula (V.57) will look like:

$$|v_{n+1} G_{n+2} - u_{n+1} G_{n+1}| = 1 \tag{V.58}$$

Let's use the recurrence formula $G_{n+2} = G_{n+1} + G_n$ for (V.58)

$$|v_{n+1} G_{n+1} + v_{n+1} G_n - u_{n+1} G_{n+1}| = |(u_{n+1} - v_{n+1})G_{n+1} - v_{n+1} G_n| = 1 \tag{V.59}$$

From (V.57) and (V.59) we can find u_n and v_n. In total, there are 4 possible combinations (V.57) and (V.59):

$$\begin{cases} v_n G_{n+1} - u_n G_n = 1 \\ (u_{n+1} - v_{n+1})G_{n+1} - v_{n+1} G_n = 1 \end{cases} \tag{V.60}$$

$$\begin{cases} v_n G_{n+1} - u_n G_n = -1 \\ (u_{n+1} - v_{n+1})G_{n+1} - v_{n+1} G_n = 1 \end{cases} \tag{V.61}$$

$$\begin{cases} v_n G_{n+1} - u_n G_n = 1 \\ (u_{n+1} - v_{n+1})G_{n+1} - v_{n+1} G_n = -1 \end{cases} \tag{V.62}$$

$$\begin{cases} v_n G_{n+1} - u_n G_n = -1 \\ (u_{n+1} - v_{n+1})G_{n+1} - v_{n+1} G_n = -1 \end{cases} \tag{V.63}$$

From (V.60) follows:

$$v_n = u_{n+1} - v_{n+1}$$
$$u_n = v_{n+1}$$
(V.64)

From the last formula follows $u_{n+1} = v_{n+2}$. Let's substitute u_{n+1} in (V.64):

$$v_n = v_{n+2} - v_{n+1} \quad \text{or} \quad v_{n+2} = v_{n+1} + v_n$$

Then v_n and u_n are recursive sequences and (V.57) will look like:

$$|v_n G_{n+1} - v_{n+1} G_n| = 1$$
(V.65)

It is obvious that from (V.63), it follows the same result.
Let's consider (V.61) and (V.62):

$$-v_n = u_{n+1} - v_{n+1} \quad \text{and} \quad -u_n = v_{n+1}$$

Then:

$$v_{n+2} = -v_{n+1} + v_n$$
(V.66)

The terms of the recurrent series from (V.66) may be negative numbers. This contradicts the condition of the fundamental theorem of phyllotaxis. Therefore, only the combinations (V.60) and (V.63) are possible.

The similarity of formulas (V.47) and (V.65) immediately strikes the eye. For their full identity, it is required to prove:

$$\text{if } |A_n G_{n+1} - A_{n+1} G_n| = 1, \text{ then } |A_1 G_2 - A_2 G_1| = 1.$$

In order to prove this, let's use the well-known formula (II.6):

$$G_n = F_{n-2} G_2 - F_{n-3} G_1$$
$$A_n = F_{n-2} A_2 - F_{n-3} A_1$$

Then:

$$|A_n G_{n+1} - A_{n+1} G_n| =$$
$$= |(F_{n-2} A_2 - F_{n-3} A_1)(F_{n-1} G_2 - F_{n-2} G_1) - (F_{n-1} A_2 - F_{n-2} A_1)(F_{n-2} G_2 - F_{n-3} G_1)| =$$
$$= |-F_{n-2} A_2 F_{n-2} G_1 - F_{n-3} A_1 F_{n-1} G_2 + F_{n-1} A_2 F_{n-3} G_1 + F_{n-2} A_1 F_{n-2} G_2| =$$
$$= |(A_2 G_1 - A_1 G_2)(F_{n-1} F_{n-3} - F_{n-2}^2)|$$

From the theory of the Fibonacci numbers, we know the formula $F_{n+1}F_{n-1} - F_n^2 = (-1)^n$, which the reader can prove themselves using Binet's formula (II.4). Consequently,

$$|A_n G_{n+1} - A_{n+1} G_n| = |A_1 G_2 - A_2 G_1| = 1$$

and $v_n = A_n$, $v_{n+1} = A_{n+1}$.

According to the fundamental theorem of phyllotaxis, the divergence coefficient lies in the interval $\left[\dfrac{A_{n+1}}{G_{n+1}}, \dfrac{A_n}{G_n}\right]$.

With increasing n, the boundaries of this interval will narrow and the divergence coefficient can be calculated as:

$$\beta = \lim_{n \to \infty} \frac{A_n}{G_n}$$

This allows us to formulate the converse of the fundamental theorem of phyllotaxis:

If the phyllotactic lattice (pattern) is generated by the non-multiple recurrent sequence with the initial terms G_1 and G_2 ($G_1, G_2 \in N$, $1 < G_1$, $2G_1 \leq G_2$; $\gcd(G_1, G_2) = 1$), then the divergence coefficient can be calculated from the system:

$$\begin{cases} |A_1 G_2 - A_2 G_1| = 1 \\ \beta = \dfrac{A_2 + \dfrac{A_1}{\tau}}{G_2 + \dfrac{G_1}{\tau}} \end{cases}, \text{ where } A_1, A_2 \in N$$

It should be noted that the encyclic numbers v and u, from the fundamental theorem of phyllotaxis, are two consecutive members of the recurrent sequence A_n. As well as A_n is the rounded to integer number of turns of the radius vector along the main genetic spiral between two nearby nodes belonging to the same thread of the n-th family. It is necessary to pay tribute to the scientific intuition of Jean, who called these numbers *encyclic* a quarter of a century ago.

V.12. THE PLANAR DH-MODEL FOR "MULTIJUGATE" SEQUENCES

Let's consider phyllotactic lattices generated by generalized recurrent sequences

$$jG_1, jG_2, jG_2 + jG_1, 2jG_2 + jG_1, 3jG_2 + 2jG_1, 5jG_2 + 3jG_1, \ldots, \quad (V.67)$$

where $1 \leq G_1$, $2G_1 \leq G_2$, $1 \leq j$, $\gcd(G_2, G_1) = 1$, $G_1, G_2, j \in N$. These recurrent sequences include all allowed generating sequences (III.1).

In [3], phyllotactic patterns generated by generalized recurrent sequence with $j \neq 1$ are called "Multijugate" sequence ("bijugate", "trijugate", "quadrijugate") and are defined as $j\langle G_1, G_2, G_2 + G_1, 2G_2 + G_1, 3G_2 + 2G_1, 5G_2 + 3G_1\rangle$. We will call such phyllotaxis as a multi-pair (two-pair, three-pair, …).

Suppose that phyllotactic lattices of multi-pair ($j > 1$) are described by the DH-Model (Section IV.2) with one pair of genetic spirals, similarly to phyllotactic lattices with $j = 1$ examined above. In the pattern generated by the sequence (V.67), the n-th parastichy family must contain jG_n different parastichies. From the Bravais-Bravais theorem, it follows that the nodes M and $M + jG_n$ will be nearby and belong to the same thread. Let's find the divergence angle $2\pi\delta$ for such the phyllotactic lattice. Similarly to Section V.2.1, the number of complete turns of the radius-vector from the node M to the point M' (Figure V.1) along the main genetic spiral is denoted as A_n, and by the ancillary genetic spiral B_n. The numbers A_n and B_n are integers because they figure out the number of complete rotations around the center Therefore the angle $|\Delta\theta|$ can be calculated in two ways:

$$|\Delta\theta| = |\Theta_m - 2\pi A_n| = 2\pi |j\delta G_n - A_n|$$

$$|\Delta\theta| = |2\pi B_n - \Theta_a| = 2\pi |B_n - j(1-\delta)G_n|$$

Because $|\Delta\theta| \to 0$, then:

$$A_n \approx j\delta G_n \quad (V.68)$$

$$B_n \approx j(1-\delta)G_n \quad (V.69)$$

It follows that A_n and B_n are members of some integral recurrent series. Moreover, $A_n < B_n$ because $0 < \delta < 0.5$. If we sum (V.68) and (V.69) we get

$$jG_n = A_n + B_n$$

From (V.68) and (II.8) follows:

$$\delta = \lim_{n \to \infty} \frac{1}{j} \frac{A_n}{G_n} = \frac{1}{j} \frac{A_2 + \dfrac{A_1}{\tau}}{G_2 + \dfrac{G_1}{\tau}} = \frac{\beta}{j}, \quad (V.70)$$

where $\beta = \dfrac{A_2 + \dfrac{A_1}{\tau}}{G_2 + \dfrac{G_1}{\tau}}$.

Let's create the phyllotactic lattice of "two-pair", "three-pair", and "four-pair" calculated according to the formula (V.70).

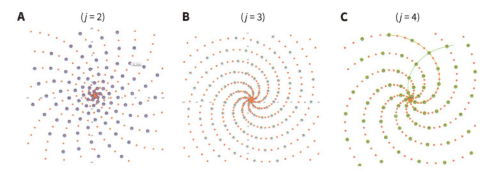

Figure V.76. Multi-pair phyllotactic lattices created according to the formula (V.70).

Let me remind you that the main criterion for the compliance of the model with real patterns is visual perception. Figure V.76 clearly shows that lattices are empty and do not have the jG_n different parastichy in each n-th family for any $n \in N$. Therefore, the DH-Model of the phyllotactic lattice generated by the generalized recurrent series for $j \neq 1$ is more complex than $j = 1$.

Jean [3] proposed to fill these lattices with nodes formed by another $j - 1$ pair of the same genetic spirals, which rotated relative to each other at an angle of $\dfrac{2\pi}{j}$. In total, there will be j pairs of genetic spirals:

$$\begin{cases} x_{m,k}(\theta) = \theta \cos\left(2\pi\left(\dfrac{\beta}{j}\theta + \dfrac{1}{j}k\right)\right) \\ y_{m,k}(\theta) = \theta \sin\left(2\pi\left(\dfrac{\beta}{j}\theta + \dfrac{1}{j}k\right)\right) \end{cases}$$

$$\begin{cases} x_{a,k}(\theta) = \theta \cos\left(2\pi\left(\left(1-\dfrac{\beta}{j}\right)\theta + \dfrac{1}{j}k\right)\right) \\ y_{a,k}(\theta) = \theta \sin\left(2\pi\left(\left(1-\dfrac{\beta}{j}\right)\theta + \dfrac{1}{j}k\right)\right) \end{cases}, \quad \text{(V.71)}$$

where $k \in N$ and takes all the values in the interval $0 \le k < j$.

It should be noted that the nodes are formed only by the intersection within each pair of genetic spirals and are not formed by the intersection of genetic spirals belonging to the different pairs. Also, multi-pair phyllotaxis does not have end-to-end nodes numbering. The nodes generated by each pair of genetic spirals have their own numbering. For example, each pair of genetic spirals has a node with the number M, which is at a distance M from the center of the lattice. Therefore, when we say *node M*, we will mean one of the nodes with the number M. The coordinates of the nodes of a multi-pair phyllotaxis will look like:

$$\begin{cases} x_k(M) = M\cos\left(2\pi\left(\dfrac{\beta}{j}M + \dfrac{1}{j}k\right)\right) \\ y_k(M) = M\sin\left(2\pi\left(\dfrac{\beta}{j}M + \dfrac{1}{j}k\right)\right) \end{cases}, \quad \text{where } 0 \le k < j,\ k, M \in N \qquad (V.72)$$

Figure V.77 shows the structure of the three-pair phyllotactic lattice. The lattice contains three pairs of genetic spirals. Each of the pairs of genetic spirals forms nodes with the divergence angle (V.70). Figure V.77.A shows the nodes formed by the first pair of genetic spirals, and Figure V.77.B and Figure V.77.C show the nodes of the other two pairs, respectively. Figure V.77.D depicts all the nodes generated by three pairs of genetic spirals.

It should be especially noted that in the DH-Model of the phyllotactic lattice with j pairs of genetic spirals, the adjacent nodes are M and $M + G_n$, and not M and $M + jG_n$, as it was assumed earlier.

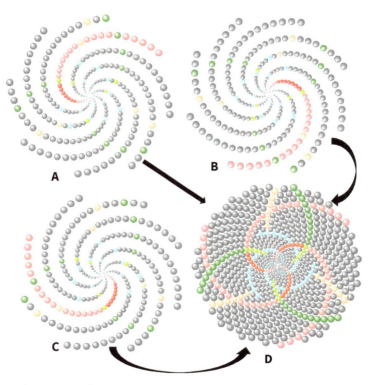

Figure V.77. The structure of the "multijugate" phyllotactic lattice.

A spiral can be drawn through the nodes of each thread of the *n*-th family. As stated above, in every *n*-th family, there will be jG_n threads:

$$\begin{cases} x(\theta) = \theta\cos\left(2\pi\left(\dfrac{(-1)^n}{G_n \tau^{n-1}} \dfrac{\beta}{j}\theta + \dfrac{1}{jG_n}s\right)\right) \\ y(\theta) = \theta\sin\left(2\pi\left(\dfrac{(-1)^n}{G_n \tau^{n-1}} \dfrac{\beta}{j}\theta + \dfrac{1}{jG_n}s\right)\right) \end{cases} \quad \text{(V.73)}$$

where $s \in \mathbb{N}$ and takes all values in the interval $0 \le s < jG_n$.

Comparing formulas (V.44) and (V.73), we can conclude that the angle $\Delta\theta_n$ (V.20) between adjacent nodes belonging to the same strand of the *n*-th degree will be equal to

$$|\Delta\theta_n| \approx \frac{2\pi}{j\sqrt{5}G_n} \quad \text{(V.74)}$$

For large M, we can take $M \gg G_n$, then the approximate ΔT value can be calculated as:

$$\Delta T \approx \sqrt{G_n^2 + \left(\frac{2\pi}{\sqrt{5}}\right)^2 \frac{M^2}{jG_n^2}} \quad \text{(V.75)}$$

Just as in Section V.5, we will find the borders of the *n*-th diapason

$$G_{n+1}^2 + \left(\frac{2\pi}{\sqrt{5}}\right)^2 \frac{1}{j^2 G_{n+1}^2} D_n^2 = G_n^2 + \left(\frac{2\pi}{\sqrt{5}}\right)^2 \frac{1}{j^2 G_n^2} D_n^2$$

$$\frac{1}{j^2}\left(\frac{2\pi}{\sqrt{5}}\right)^2 D_n^2 \left(\frac{1}{G_n^2} - \frac{1}{G_{n+1}^2}\right) = G_{n+1}^2 - G_n^2$$

$$D_n = \frac{\sqrt{5}}{2\pi} jG_{n+1}G_n \approx \frac{\tau^n}{2\pi\beta} jG_n \quad \text{(V.76)}$$

and nodes of rise

$$G_{n-1}^2 + \left(\frac{2\pi}{\sqrt{5}}\right)^2 \frac{W_n^2}{j^2 G_{n-1}^2} = G_{n+1}^2 + \left(\frac{2\pi}{\sqrt{5}}\right)^2 \frac{W_n^2}{j^2 G_{n+1}^2}$$

$$\frac{1}{j^2}\left(\frac{2\pi}{\sqrt{5}}\right)^2 W_n^2 \left(\frac{1}{G_{n-1}^2} - \frac{1}{G_{n+1}^2}\right) = G_{n+1}^2 - G_{n-1}^2$$

$$W_n = \frac{\sqrt{5}}{2\pi} jG_{n-1}G_{n+1} \approx \frac{\sqrt{5}}{2\pi} jG_n^2 \qquad (V.77)$$

Just as in Section V.6, we will find the function of dependence of the radius of incompressible and deformable element of the phyllotactic lattice on the node number:

$$\begin{cases} M = \dfrac{\sqrt{5}}{2\pi} \tau j G_n^2 \\ d(M) = \sqrt{G_n^2 + \left(\dfrac{2\pi}{\sqrt{5}}\right)^2 \dfrac{M^2}{j^2 G_n^2}} \end{cases}$$

We substitute the first equality $G_n^2 = \dfrac{2\pi}{\sqrt{5}\tau} \dfrac{M}{j}$ into the second to exclude n:

$$d(M) = \sqrt{\frac{2\pi}{\sqrt{5}\tau}\frac{M}{j} + \left(\frac{2\pi}{\sqrt{5}}\right)^2 M^2 \frac{\sqrt{5}\tau}{2\pi jM}} = \sqrt{\frac{2\pi}{\sqrt{5}\tau}\frac{M}{j} + \frac{2\pi\tau}{\sqrt{5}}\frac{M}{j}} = \sqrt{2\pi\frac{M}{j}} \qquad (V.78)$$

Find the packing density for jL incompressible and deformable elements, similarly to (V.27):

$$j\sum_{i=1}^{L} \pi\left(\frac{d_i}{2}\right)^2 : \pi L^2 = \frac{\pi j}{2}\sum_{i=1}^{L}\left(\sqrt{\frac{i}{j}}\right)^2 : L^2 = \frac{\pi \dfrac{L(L-1)}{2}}{L^2} \approx \frac{\pi}{4} \qquad (V.79)$$

Same as it is in Section V.7, we will find the angle between tangent lines to the opposed parastichies, using the formula of spirals (V.52) on which the threads of the nodes of the phyllotactic lattice lie, generated by generalized recurrent sequences.

$$\gamma(M) = arctan\left(\frac{2\pi}{jG_n \tau^{n-1}} \beta M\right) + arctan\left(\frac{2\pi}{jG_{n+1} \tau^n} \beta M\right) \qquad (V.80)$$

If in (V.79) we substitute the border of the diapason from (V.76), then the angle between tangent lines to the opposed parastichy pair will be a right angle:

$$\gamma(D_n) = arctan(\tau) + arctan\left(\frac{1}{\tau}\right) = \frac{\pi}{2}$$

If we substitute $j = 1$ into formulas (V.70)–(V.80), then they will turn into the corresponding formulas for the phyllotactic lattice generated by the non-multiple sequence. This proves that the phyllotactic lattice generated by the non-multiple sequence is a particular case of the phyllotactic lattice generated by the generalized recurrent sequences.

In the figures below, you can see the summary chart for phyllotactic lattices generated by generalized recurrent sequences with different initial terms and j.

j	G_1	G_2	Figure	URL
2	1	2	Figure V.78	https://youtu.be/sHV_XJe7DUA
2	1	3	Figure V.79	https://youtu.be/izQ9rWGONzU
2	1	4	Figure V.80	https://youtu.be/lxbOBL2Sb88
2	1	5	Figure V.81	https://youtu.be/AlmVDwn4fTk
2	1	6	Figure V.82	https://youtu.be/rr7q9RY6_bI
2	1	7	Figure V.83	https://youtu.be/2p7WY40h_88
2	1	8	Figure V.84	https://youtu.be/NCENGkDVPWo
2	2	5	Figure V.85	https://youtu.be/gEsu89NKIss
2	2	7	Figure V.86	https://youtu.be/bB4vevg7Q-8
2	2	9	Figure V.87	https://youtu.be/Yfn32j0Lk9c
2	2	11	Figure V.88	https://youtu.be/D21-6_mzJkE
2	3	7	Figure V.89	https://youtu.be/gltdphc-C6k
2	3	8	Figure V.90	https://youtu.be/-V9m5iBpWhc
2	3	10	Figure V.91	https://youtu.be/hvtTiABTTUg
2	3	11	Figure V.92	https://youtu.be/BvfE9xxcvlI
2	4	9	Figure V.93	https://youtu.be/Zj80vRzyqr8
2	4	11	Figure V.94	https://youtu.be/uyz47_TsIDY
2	4	13	Figure V.95	https://youtu.be/VoRHiYUSg-Y
2	5	11	Figure V.96	https://youtu.be/H7MjlMamxBk
2	5	12	Figure V.97	https://youtu.be/74Jc42Kjg4k
2	5	13	Figure V.98	https://youtu.be/mXXpmsyJSb8
3	1	2	Figure V.99	https://youtu.be/TtFSfexak3g
3	1	3	Figure V.100	https://youtu.be/rI_aJD9zegg
3	1	4	Figure V.101	https://youtu.be/g9Xm7f03IQI
3	1	5	Figure V.102	https://youtu.be/dTZYE8W8EIk
3	1	6	Figure V.103	https://youtu.be/kxTRVopX6mg
3	2	5	Figure V.104	https://youtu.be/MmNk-7yyLE4
3	2	7	Figure V.105	https://youtu.be/Z8qNwt6lYy8
3	2	9	Figure V.106	https://youtu.be/E7k-MKhzwA4

j	G_1	G_2	Figure	URL
3	3	7	Figure V.107	https://youtu.be/U5Nbwlfor9w
3	3	8	Figure V.108	https://youtu.be/y5X10THbtS0
3	4	9	Figure V.109	https://youtu.be/rVh_Majd12Q
3	5	11	Figure V.110	https://youtu.be/GxUHIpYGuIM
3	5	12	Figure V.111	https://youtu.be/1i0RJpzvQvc
4	1	2	Figure V.112	https://youtu.be/XSGNHPd9k5Y
4	1	3	Figure V.113	https://youtu.be/4RBLZXtGOmM
4	1	4	Figure V.114	https://youtu.be/ht_Kf8EYBvc
4	2	5	Figure V.115	https://youtu.be/YfTNyFKLP6Y
4	3	7	Figure V.116	https://youtu.be/M4XuW2GsB9A
5	1	2	Figure V.117	https://youtu.be/pKcrUdF2vOA
5	1	3	Figure V.118	https://youtu.be/jAF1nWKiwm0
5	2	5	Figure V.119	https://youtu.be/bVg62Lf7u0s

V.13. THE UNIVERSAL ALGORITHM FOR CALCULATING THE DIVERGENCE COEFFICIENT

A universal algorithm for calculating the coefficient of discrepancy follows from Converse of the fundamental theorem of phyllotaxis for generalized recurrent sequences.

If the lattice (pattern) of phyllotaxis is generated by a generalized recurrent next to the initial terms jG_1 and jG_2 ($j, G_1, G_2 \in N$, $1 \leq G_1$, $2G_1 \leq G_2$; $\gcd(G_1, G_2) = 1$), then the divergence coefficient can be calculated from the system:

$$\begin{cases} |A_1 G_2 - A_2 G_1| = 1 \\ \beta = \dfrac{1}{j} \dfrac{A_2 + \dfrac{A_1}{\tau}}{G_2 + \dfrac{G_1}{\tau}} \end{cases}, \text{ where } A_1, A_2 \in N$$

The universal algorithm for calculating the divergence coefficient allows you to draw the phyllotactic lattice for any of the most exotic opposed pair.

Let's draw a lattice (pattern) phyllotaxis containing a visible opposed parastichy pair (L, K), where $K > L \geq 1$, $L, K \in N$:

1. If (L, K), the index of opposed parastichy pair, then the numbers L and K are neighboring members of some generalized recurrent sequence $jG_1, jG_2, jG_2 + jG_1$, $2jG_2 + jG_1, 3jG_2 + 2jG_1, 5jG_2 + 3jG_1, \ldots$,

- If L and K have a common divisor, then $j = \gcd(L, K)$; if not, then $j = 1$,
- Find G_1 and G_2, producing successive subtractions. $(K/j - L/j)$, $(L/j - (K/j - L/j))$, $((K/j - L/j) - (L/j - (K/j - L/j)))$, ..., until the last two results of subtraction satisfy the condition $1 \leq G_1, 2G_1 \leq G_2$,

2. Find A_1 and A_2 as the smaller solution of the Diophantine equation $|A_1 G_2 - A_2 G_1| = 1$, using the Euclidean algorithm [17],
3. Calculate the divergence coefficient according to the formula $\beta = \dfrac{1}{j} \dfrac{A_2 + \dfrac{A_1}{\tau}}{G_2 + \dfrac{G_1}{\tau}}$.

Let's us explain the universal algorithm for calculating the divergence coefficient in two examples:

Example #1. Opposed parastichy pair (29, 46).

1. Numbers 29 and 46 do not have a common divisor, so, $j = 1$. Let's find the initial terms: $46 - 29 = 17$, $29 - 17 = 12$, $17 - 12 = 5$. Then, $G_1 = 5$ and $G_2 = 12$.
2. The solutions of the integer equation $|12A_1 - 5A_2| = 1$ will be pairs (2, 5) and (3, 7). The smaller pair is $A_1 = 2$, $A_2 = 5$.
3. Let's calculate $\beta = \dfrac{1}{1} \dfrac{5 + \dfrac{2}{\tau}}{12 + \dfrac{5}{\tau}} = 0.41323661$, $2\pi\beta = 148.771318°$

A combined chart for this phyllotactic lattice can be seen in the table above Figure V.52.

Example #2. Opposed parastichy pair (28, 46).

1. The greatest multiple divider of numbers 28 and 46 is 2, so, $j = 2$. Find the initial terms: $46/2 - 28/2 = 9$, $28/2 - 9 = 5$, $9 - 5 = 4$, $5 - 4 = 1$. Then, $G_1 = 1$ and $G_2 = 5$.
2. The solutions of the integer equation $|5A_1 - 1A_2| = 1$ will be pairs (0, 1) and (1, 4). Then, $A_1 = 0$, $A_2 = 1$.
3. Let's calculate $\beta = \dfrac{1}{2} \dfrac{1 + \dfrac{0}{\tau}}{5 + \dfrac{1}{\tau}} = 0.088999106$, $2\pi\beta = 32.039678°$

A combined chart for this phyllotactic lattice can be seen in the above Figure V.81.

V.14. INTERIM RESULTS

The above analysis showed a complete correspondence of the DH-Model to phyllotactic patterns on the planar Archimedean spirals, to the Bravais-Bravais theorem, and to the fundamental theorem of phyllotaxis. This allows us safely to conclude that the DH-Model is adequate to the real botanical objects.

The DH-Model analysis allowed to obtain analytically the previously known, but empirically found, numerical relations:

- the angle between tangent lines to the parastichies is equal to a right angle in the zone of the best visibility of opposed parastichy pair. This is derived from a formula for calculating the angle between tangent lines to parastichies;
- a new proof of the fundamental theorem of phyllotaxis and finding out the meaning of the encyclic numbers u and v.

Also, on the basis of the DH-Model, there were also analytically found previously unknown numerical ratios:

- the universal algorithm for calculating the divergence coefficient for the phyllotactic lattice, which is generated by a recursive sequence with any allowed initial terms;
- a one-to-one correspondence between the divergence angle and the generating recursive sequence;
- the formula calculating the border of full and partial visibility of parastichies;
- the formula calculating the node of "the phyllotaxis rises";
- the formula calculating the diameter of the primordium, from which follows that the diameter does not depend on the angle of divergence or the generating recurrent sequence;
- the packing density of elements is a constant for the phyllotactic lattice on the planar Archimedean spirals and does not depend on the angle of divergence or the generating recurrent sequence.

From the previous above, it follows that the Fibonacci sequence, as generating sequence of the phyllotactic lattice, does not have "advantages" compared to the other recurrent sequences. This contradicts to the observed distribution of phyllotactic patterns in nature.

FIBONACCI LATTICES ON THE PLANAR NON-ARCHIMEDEAN SPIRALS

Let us verify the theoretical possibility of constructing Fibonacci lattices on the non-Archimedean genetic spirals. From the definition of genetic spirals in (IV.3) and (IV.4), it follows that the divergence coefficient and Edge function are independent parameters of the phyllotactic lattice. Therefore, the numerical relations for the angular dimensions of the lattice found for the Archimedean phyllotaxis will also be true for the non-Archimedean phyllotaxis, which allows us to restrict ourselves to considering Fibonacci phyllotaxis on the non-Archimedean spirals.

In the non-Archimedean Fibonacci phyllotaxis, genetic spirals are described by a parametric system:

$$\begin{cases} x_m(\theta) = E(\theta) \cdot \cos\left(\frac{2\pi}{\tau^2}\theta\right) \\ y_m(\theta) = E(\theta) \cdot \sin\left(\frac{2\pi}{\tau^2}\theta\right) \end{cases} \quad \text{(VI.1)}$$

$$\begin{cases} x_a(\theta) = E(\theta) \cdot \cos\left(-\frac{2\pi}{\tau}\theta\right) \\ y_a(\theta) = E(\theta) \cdot \sin\left(-\frac{2\pi}{\tau}\theta\right) \end{cases} \quad \text{(VI.2)}$$

where Edge function $E(\theta)$ is a real continuous monotonically increasing function. From the entire variety of Edge functions, it suffices to consider the exponential $E(\theta) = w^\theta = e^{\ln(w)\theta}$ and the power $E(\theta) = \theta^v$ functions, since numerous scientific works believe that they most closely describe phyllotaxis in biological objects.

To eliminate confusion, the lattice node number of the non-Archimedean Fibonacci phyllotaxis will be denoted by m:

$$\begin{cases} x(m) = E(m) \cdot \cos\left(\frac{2\pi}{\tau^2}m\right) \\ y(m) = E(m) \cdot \sin\left(\frac{2\pi}{\tau^2}m\right) \end{cases}$$

If we recall that if the Edge function is the exponent $E(m) = w^m$, then both the spiral and phyllotaxis are usually called logarithmic. If the Edge function is the exponential function $E(m) = m^v$, then we will call them *power spiral* and *power phyllotaxis*.

Let us return to the Figure V.1 and consider the KMO triangle for the genetic spirals (VI.1) and (VI.2). In order to get Fibonacci phyllotaxis, it is necessary that the nodes K and M to belong to the same thread and be nearby:

$$M - K = F_n$$

Then (V.21), for the non-Archimedean phyllotaxis, will look:

$$\Delta T^2 \approx (E(m) - E(m - F_n))^2 + E(m)E(m - F_n)(\Delta\theta)^2 \tag{VI.3}$$

The angle between two nearby nodes M and K will be the same as in (V.20):

$$|\Delta\theta| \approx \frac{2\pi}{\sqrt{5}} \frac{1}{F_n} \tag{VI.4}$$

Let's substitute (VI.4) in (VI.3):

$$\Delta T = \sqrt{(E(m) - E(m - F_n))^2 + E(m)E(m - F_n)\left(\frac{2\pi}{\sqrt{5}} \frac{1}{F_n}\right)^2} \tag{VI.5}$$

Due to the conservation of angular relations, the nodes of lattice of the non-Archimedean phyllotaxis form threads. Nodes belonging to one thread from the n-th thread family will have numbers

$$s + i \cdot F_n$$

and lie on a spiral:

$$\begin{cases} x(\theta) = E(\theta)\cos\left(2\pi\left(\frac{(-1)^n}{F_n \tau^{n+1}}\theta + \frac{1}{F_n}s\right)\right) \\ y(\theta) = E(\theta)\sin\left(2\pi\left(\frac{(-1)^n}{F_n \tau^{n+1}}\theta + \frac{1}{F_n}s\right)\right) \end{cases}$$

VI.1. FIBONACCI LATTICES ON THE PLANAR POWER GENETIC SPIRALS

As mentioned in Section II.3, for the power spirals, the Edge function will be $E(m) = m^v$. Therefore, the distance between nearby nodes (VI.5) for the power phyllotaxis is

$$\Delta T = \sqrt{(m^v - (m-F_n)^v)^2 + m^v(m-F_n)^v \left(\frac{2\pi}{\sqrt{5}}\frac{1}{F_n}\right)^2}$$

Let us use the approximate calculation formula for $\Delta x \to 0$:

$$(1+\Delta x)^v \approx 1 + v \cdot \Delta x$$

Given that $m \gg F_n$:

$$(m-F_n)^v = m^v\left(1-\frac{F_n}{m}\right)^v \approx m^v\left(1-\frac{vF_n}{m}\right) = m^v - m^{v-1}vF_n$$

Consequently,

$$\Delta T \approx \sqrt{(m^{v-1}vF_n)^2 + m^{2v}\left(\frac{2\pi}{\sqrt{5}}\frac{1}{F_n}\right)^2} = m^{v-1}\sqrt{v^2F_n^2 + m^2\left(\frac{2\pi}{\sqrt{5}}\frac{1}{F_n}\right)^2} \quad (VI.6)$$

As in the Archimedean phyllotaxis, the distance between two nearby nodes of the power phyllotaxis depends on the node number m and the degree of the thread family n, to which these nodes belong. Note that for $v = 1$, formula (VI.6) turns into (V.22).

Let's find the boundaries of the ranges in the same way as in (V.23):

$$v^2F_{n+1}^2 + \left(\frac{2\pi}{\sqrt{5}}\right)^2 \frac{D_n^2}{F_{n+1}^2} = v^2F_n^2 + \left(\frac{2\pi}{\sqrt{5}}\right)^2 \frac{D_n^2}{F_n^2}$$

$$D_n = \frac{\sqrt{5}}{2\pi}vF_{n+1}F_n \approx \frac{\sqrt{5}\tau}{2\pi}vF_n^2$$

Let's find the nodes of "the phyllotaxis rises" the same way as in (V.25):

$$v^2F_{n-1}^2 + \left(\frac{2\pi}{\sqrt{5}}\right)^2 \frac{W_n^2}{F_{n-1}^2} = v^2F_{n+1}^2 + \left(\frac{2\pi}{\sqrt{5}}\right)^2 \frac{W_n^2}{F_{n+1}^2}$$

$$W_n = \frac{\sqrt{5}}{2\pi}vF_{n-1}F_{n+1} \approx \frac{\sqrt{5}}{2\pi}vF_n^2$$

Let's find the dependence function of the radius of incompressible and deformable element of the phyllotactic lattice the same way as in (V.26):

$$\begin{cases} m = \dfrac{\sqrt{5}}{2\pi} \tau v F_n^2 \\ d(M) = \sqrt{v^2 F_n^2 + \left(\dfrac{2\pi}{\sqrt{5}}\right)^2 \dfrac{M^2}{F_n^2}} \end{cases}$$

Let's exclude n, by substituting $F_n^2 = \dfrac{2\pi}{\sqrt{5}\tau v} m$, then

$$d(m) = m^{v-1}\sqrt{\dfrac{2\pi}{\sqrt{5}\tau} vm + \left(\dfrac{2\pi}{\sqrt{5}}\right)^2 m^2 \dfrac{\sqrt{5}\tau v}{2\pi m}} = m^{v-1}\sqrt{\dfrac{2\pi}{\sqrt{5}\tau} vm + \dfrac{2\pi\tau}{\sqrt{5}} vm} = \sqrt{2\pi v} m^{v-0.5}$$

(VI.7)

If $v = 0.5$ in (VI.7), then the diameter of the disks becomes constant. This confirms the well-known fact that in the phyllotactic lattice built on a parabolic spiral $r = \theta^{\frac{1}{2}}$, all primordia have the same diameter. However, confirmation of the existence of real inflorescences with the same primordia requires more accurate measurements.

Let's find the packing density for L incompressible and deformable elements as the ratio of the sum of the disk areas to the area of the entire inflorescence. If L is the node number, then the distance from the center of the lattice to this node is L^v, and the diameter of the disk with number L is $\sqrt{2v\pi}L^{v-0.5}$.

$$\sum_{i=0}^{L} \pi \left(\dfrac{1}{2}\sqrt{2\pi v} i^{v-0.5}\right)^2 : \pi L^{2v} = \dfrac{\pi v}{2} \sum_{i=0}^{L} (i^{2v-1}) : L^{2v} \qquad \text{(VI.8)}$$

Let's use Faulhaber's formula:

$$\sum_{i=0}^{L} i^p \approx \dfrac{L^{p+1}}{p+1}$$

Then:

$$\sum_{i=0}^{L} i^{2v-1} \approx \dfrac{L^{2v}}{2v} \qquad \text{(VI.9)}$$

Let's substitute (VI.9) into (VI.8):

$$\dfrac{\pi v}{2}\sum_{i=0}^{L}(i^{2v-1}) : L^{2v} \approx \dfrac{\pi v}{2} \dfrac{L^{2v}}{2v} \dfrac{1}{L^{2v}} = \dfrac{\pi}{4}$$

As it can be seen from the last formula, the density of the lattice filling of power phyllotaxis does not depend on v and completely coincides with the results obtained for phyllotaxis on Archimedean spirals (V.27).

Let's find the angle between the tangents to the opposed parastichy pair for the lattice of the power phyllotaxis, similarly to Section IV.7.

The nodes belonging to one thread with degree n will lie on a spiral:

$$\begin{cases} x(\theta) = \theta^v \cos\left(2\pi\left(\dfrac{(-1)^n}{F_n \tau^{n+1}}\theta + \dfrac{1}{F_n}s\right)\right) \\ y(\theta) = \theta^v \sin\left(2\pi\left(\dfrac{(-1)^n}{F_n \tau^{n+1}}\theta + \dfrac{1}{F_n}s\right)\right) \end{cases} \quad \text{(VI.10)}$$

Moreover, for $n = 1$ (VI.10), it becomes the main genetic spiral, and for $n = 0$ is an ancillary.

Accordingly, the nodes belonging to another thread with degree $n + 1$ will lie on a spiral:

$$\begin{cases} x(\theta) = \theta^v \cos\left(2\pi\left(\dfrac{(-1)^{n+1}}{F_{n+1} \tau^{n+2}}\theta + \dfrac{1}{F_{n+1}}s\right)\right) \\ y(\theta) = \theta^v \sin\left(2\pi\left(\dfrac{(-1)^n}{F_{n+1} \tau^{n+2}}\theta + \dfrac{1}{F_{n+1}}s\right)\right) \end{cases}$$

From analytic geometry (Appendix D, formula D.7), it is known that the tangent function of the angle between the tangent and the radius vector, which is passing through the tangent point, for a power spiral is:

$$\tan(\gamma_n) = \left|\dfrac{2\pi}{v}\dfrac{(-1)^n}{F_n \tau^{n+1}}\theta\right| \qquad \tan(\gamma_{n+1}) = \left|\dfrac{2\pi}{v}\dfrac{(-1)^{n+1}}{F_{n+1} \tau^{n+2}}\theta\right|$$

Then

$$\gamma(m) = \gamma_n + \gamma_{n+1} = \arctan\left(\dfrac{2\pi m}{vF_n \tau^{n+1}}\right) + \arctan\left(\dfrac{2\pi m}{vF_{n+1} \tau^{n+2}}\right)$$

Let's find the angle between the tangents to the threads of the n-th and the $(n+1)$-th degree for the node D_n, which is the upper boundary of the n-th diapason:

$$\gamma(D_n) = \arctan(\tau) + \arctan\left(\dfrac{1}{\tau}\right) = \dfrac{\pi}{2}$$

It turns out that for power phyllotaxis, the tangents to the opposed parastichies intersect at a right angle in the zone of the best visibility.

Let's find the "Plastochrone ratio R" for the lattice of the power phyllotaxis:

$$R = \frac{E(m+1)}{E(m)} = \frac{(m+1)^v}{m^v} = \left(1 + \frac{1}{m}\right)^v$$

$$\log(R) = \log\left(1 + \frac{1}{m}\right)^v = v\log\left(1 + \frac{1}{m}\right)$$

Just as for the Archimedean phyllotaxis, the "Plastochrone ratio R" for the power phyllotaxis can be neither a constant not a linear function and does not depend on the divergence coefficient.

Just like the lattice of the Archimedean phyllotaxis, the lattice of the power phyllotaxis will have structural recursiveness

$$\begin{cases} x(m+1) = \left(1 + \sqrt[2v]{x^2(m) + y^2(m)}\right)^v \cos\left(2\pi\beta + \arctg\left(\frac{y(m)}{x(m)}\right)\right) \\ y(m+1) = \left(1 + \sqrt[2v]{x^2(m) + y^2(m)}\right)^v \sin\left(2\pi\beta + \arctg\left(\frac{y(m)}{x(m)}\right)\right) \\ dr(m+1) = \sqrt{2\pi\left(1 + \sqrt[2v]{x^2(m) + y^2(m)}\right)^v} \end{cases}$$

but will not have structural self-similarity.

Video VI.1[1] shows how the pattern of the power phyllotaxis, graphs of the distance between nearby nodes and the angle between tangents to the threads change, when altering the parameter $v = 0.3 \div 3$.

The patterns of power phyllotaxis in the range $v = 0.3 \div 0.5$ are of only theoretical interest, because the growth function of primordia (VI.7) should be increasing. At $v = 0.5$, all primordia are the same size, and at $v = 1$, we get the Archimedean phyllotaxis already studied above. It should be noted that the patterns of power phyllotaxis at $v = 1/\tau$ or $v = \tau$ do not have visible qualitative differences from the rest of the patterns.

VI.2. FIBONACCI LATTICES ON THE PLANAR LOGARITHMIC GENETIC SPIRALS

As mentioned in Section II.3, for logarithmic spirals, the Edge function will be $E(m) = w^m$. Therefore, the distance between nearby nodes (VI.5) for logarithmic phyllotaxis is

---

[1] The video is located at https://youtu.be/rWrj_1P6gLU

$$\Delta T = \sqrt{(w^m - w^{m-F_n})^2 + w^m \, w^{m-F_n} \left(\frac{2\pi}{\sqrt{5}} \frac{1}{F_n}\right)^2}$$

Given that $m \gg F_n$:

$$\Delta T \approx w^m \sqrt{(1 - w^{-F_n})^2 + w^{-F_n} \left(\frac{2\pi}{\sqrt{5}} \frac{1}{F_n}\right)^2} \qquad (VI.11)$$

Formula (VI.11) is interesting in that the expression under the square root does not contain the node number m, i.e. is a constant. That will mean that the distance between nearby nodes (VI.11) is equal to the exponential function w^m multiplied by the constant.

Let's find the boundaries of the ranges in the same way as in (V.23):

$$w^{2m}\left((1 - w^{-F_n})^2 + w^{-F_n}\left(\frac{2\pi}{\sqrt{5}}\frac{1}{F_n}\right)^2\right) = w^{2m}\left((1 - w^{-F_{n+1}})^2 + w^{-F_n}\left(\frac{2\pi}{\sqrt{5}}\frac{1}{F_{n+1}}\right)^2\right) \qquad (VI.12)$$

It is seen from (VI.12) that there is no m satisfying this equation; therefore, there are no diapasons of the best visibility and the nodes of "the phyllotaxis rises" on logarithmic spiral lattices.

Video VI.2² shows the patterns of logarithmic phyllotaxis when changing the parameter $w = 1.00005 \div 1.03$. As it can be seen from this video, the parastichies are reliably visible only in the fairly narrow range $w = 1.00045 \div 1.005$. The graphs of the distance between the neighboring nodes of the threads do not intersect for any w, i.e. there is no visual effect of "rise" in the logarithmic phyllotaxis. Also, the pattern of logarithmic phyllotaxis has in the center a "blind spot", which we do not find in the real inflorescences.

Let's find the angle between the tangents to the opposed parastichies for the lattice of the logarithmic phyllotaxis similarly to Section IV.7.

The nodes belonging to one thread with degree n will lie on a spiral:

$$\begin{cases} x(\theta) = w^\theta \cos\left(2\pi\left(\frac{(-1)^n}{F_n \tau^{n+1}}\theta + \frac{1}{F_n}s\right)\right) \\ y(\theta) = w^\theta \sin\left(2\pi\left(\frac{(-1)^n}{F_n \tau^{n+1}}\theta + \frac{1}{F_n}s\right)\right) \end{cases} \qquad (VI.13)$$

---

² The video is located at https://youtu.be/GRTCG1KC2EU

Moreover, for $n = 1$ (VI.13), it becomes the main genetic spiral, and for $n = 0$ is an ancillary.

Accordingly, the nodes belonging to another thread with degree $n + 1$ will lie on the spiral:

$$\begin{cases} x(\theta) = w^{\theta} \cos\left(2\pi\left(\frac{(-1)^{n+1}}{F_{n+1}\tau^{n+2}}\theta + \frac{1}{F_{n+1}}s\right)\right) \\ y(\theta) = w^{\theta} \sin\left(2\pi\left(\frac{(-1)^{n}}{F_{n+1}\tau^{n+2}}\theta + \frac{1}{F_{n+1}}s\right)\right) \end{cases}$$

From analytic geometry (Appendix **D**, formula (D.8)), it is known that the tangent function of the angle between the tangent and the radius vector, passing through the tangent point, for a power spiral is:

$$\tan(\gamma_n) = \left|\frac{2\pi}{\ln(w)}\frac{(-1)^n}{F_n\tau^{n+1}}\right| \qquad \tan(\gamma_{n+1}) = \left|\frac{2\pi}{\ln(w)}\frac{(-1)^{n+1}}{F_{n+1}\tau^{n+2}}\right|$$

Then:

$$\gamma(M) = \gamma_n + \gamma_{n+1} = \arctan\left(\frac{2\pi}{\ln(w)F_n\tau^{n+1}}\right) + \arctan\left(\frac{2\pi}{\ln(w)F_{n+1}\tau^{n+2}}\right) \quad (VI.14)$$

As it can be seen from (VI.14), the angle between the tangents to the threads of the n-th and the $(n + 1)$-th degree is a constant and, in the general case, it is not equal to a right angle.

For the logarithmic phyllotaxis, the "Plastochrone ratio R" will be constant, regardless of the divergence angle and the node number:

$$R = \frac{E(m+1)}{E(m)} = \frac{w^{m+1}}{w^m} = w$$

Unlike the phyllotactic lattice on power spirals, the phyllotactic lattice on logarithmic spirals will have both structural recursiveness and structural self-similarity, because the logarithmic spiral can transform into itself when scaling:

$$\begin{cases} x(m+1) = w\sqrt{x^2(m) + y^2(m)} \cos\left(2\pi\beta + \arctg\left(\frac{y(m)}{x(m)}\right)\right) \\ y(m+1) = w\sqrt{x^2(m) + y^2(m)} \sin\left(2\pi\beta + \arctg\left(\frac{y(m)}{x(m)}\right)\right), \\ dr(m+1) = \sqrt{2\pi w}\sqrt{x^2(m) + y^2(m)} \end{cases}$$

There is a decomposition of the function $f(x) = w^x$ in the power Maclaurin series:

$$f(x) = w^x = 1 + \frac{x\ln(w)}{1!} + \frac{(x\ln(w))^2}{2!} + \frac{(x\ln(w))^3}{3!} + \ldots + \frac{(x\ln(w))^n}{n!} + \ldots$$

On the other hand, with $w \approx 1$, $\ln(w) \approx w - 1$, then:

$$f(x) = w^x \approx 1 + x\ln(w) \approx 1 + (w-1)x \qquad \text{(VI.15)}$$

It follows from (VI.15) that at $w \approx 1$, the power function w^x becomes close to linear, and the genetic spiral (VI.13) turns from a logarithmic to an Archimedean one. Weise in [18] noted the indistinguishability of the outer turns of the logarithmic and arithmetic spirals. Indeed, it is very problematic to distinguish a logarithmic spiral with $w = 1.000255$ in Figure VI.1[3] from the Archimedean spiral. It is also very difficult to distinguish between the logarithmic and the power spirals at small values of w and v: for example, Figure VI.2[4] (logarithmic spiral $w = 1.00016$, power $v = 0.62$) and Figure VI.3[5] ($w = 1.0001$, $v = 0.37$).

VI.3. INTERIM RESULTS

The phyllotaxis model on the planar power spirals, a case of which is the Archimedean phyllotaxis, describes the well-known visual effects and numerical relationships in inflorescence of plants.

However, the DH-Model on logarithmic spirals has a number of flaws:

- There is no visual effect of "the phyllotaxis rises";
- There are no diapasons of the best visibility of the parastichies. All the parastichies are equally visible on the entire phyllotactic lattice;
- As it can be seen from (VI.14), the angle between the tangents to the threads of the n-th and the $(n + 1)$-th degree is a constant and in general case is not equal to a right angle;
- The angle between the tangents to the threads is a constant and in general case is not equal to a right angle.

What was said above, allows the author to suggest that the genetic spirals of the phyllotaxis lattice cannot be logarithmic. Weise [18], using visual-geometric analysis, argued: "The logarithmic spiral is not a good stencil for the phyllotactic pattern."

---

[3] The image is located at https://youtu.be/VTOwIzlRwNU
[4] The image is located at https://youtu.be/5LIH1khsdBQ
[5] The image is located at https://youtu.be/QJyaJSLoTPw

Almost all researchers of phyllotaxis, beginning with Church [19], were of the opinion that there are botanical objects with logarithmic phyllotaxis. According to the author, this stereotype arose because of a false analogy between phyllotaxis and spiral shells of sea snails, which are really described by logarithmic spirals, and also because of errors in phyllotaxis measuring.

However, today there is not enough data to draw an unambiguous conclusion about the existence or nonexistence of objects with a phyllotactic lattice on logarithmic spirals.

As it was shown above, it is very difficult to distinguish a logarithmic spiral with $w = 1.003 \pm 0.25\%$ from the Archimedean or from a power spiral due to the absence of an appropriate measuring tool. The principles for creating such a tool for measuring of phyllotaxis will be suggested in this study below.

THE CYLINDRICAL DH-MODEL

By analogy, with the planar DH-Model, the cylindrical DH-Model is also based on two helices, which are called the main (VII.1) and ancillary (VII.2) genetic helix:

$$\begin{cases} x_m(\theta) = R \cdot \cos(2\pi\beta\theta) \\ y_m(\theta) = R \cdot \sin(2\pi\beta\theta), \\ z_m(\theta) = H_m \cdot \beta\theta \end{cases} \quad \text{(VII.1)}$$

$$\begin{cases} x_a(\theta) = R\cos(2\pi(1-\beta)\theta) \\ y_a(\theta) = -R\sin(2\pi(1-\beta)\theta), \\ z_a(\theta) = H_a \cdot (1-\beta)\theta \end{cases} \quad \text{(VII.2)}$$

where $\theta > 0$ is independent real variable, $0 \leq \beta \leq 0.5$ is divergence coefficient, R is cylinder diameter, H_m is pitch of the main genetic helix, and H_a is pitch of the ancillary genetic helix.

The lattice nodes of the cylindrical phyllotaxis will be located at the points of mutual intersection of the genetic helices.

In the main genetic helix:

$$\begin{cases} x_m(M) = R \cdot \cos(2\pi\beta M) \\ y_m(M) = R \cdot \sin(2\pi\beta M) \\ z_m(M) = H_m \cdot \beta M \end{cases} \quad \text{(VII.3)}$$

In the ancillary genetic helix:

$$\begin{cases} x_a(M) = R \cdot \cos(-2\pi(1-\beta)M) \\ y_a(M) = R \cdot \sin(-2\pi(1-\beta)M) \\ z_a(M) = H_a \cdot (1-\beta)M \end{cases} \quad \text{(VII.4)}$$

From (VII.3) and (VII.4), it follows that H_m and H_a are interconnected via the divergence coefficient:

$$\frac{H_m}{H_a} = \frac{\beta}{1-\beta}$$

Therefore, in the future, we will denote the step of the main genetic helix as H.

Unlike the planar DH-Model, the structure of the nodes of the phyllotactic lattice in the cylindrical DH-Model has the property of isomorphism, i.e. when turning around the Z axis and shifting along the same axis, each M-th node goes into the $(M + i)$-th node, where $i \in Z$, $i \neq 0$. Therefore, the initial node of the cylindrical phyllotaxis lattice can be chosen arbitrarily.

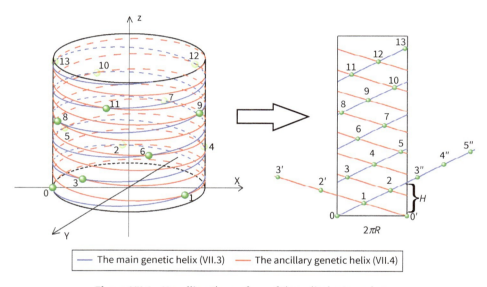

— The main genetic helix (VII.3)   — The ancillary genetic helix (VII.4)

Figure VII.1. Unrolling the surface of the cylinder to a plane.

When we unroll the surface of the cylinder along the axis of symmetry to a plane (Figure VII.1), its result is called *the planar representation of the cylindrical phyllotaxis*.

The nodes of the phyllotactic lattice will then have such coordinates in terms of a two-dimensional Cartesian coordinate system:

$$\begin{cases} x_m(M) = 2\pi R(\beta M - [\beta M]) \\ y_m(M) = H\beta M \end{cases} \quad \text{(VII.5)}$$

$$\begin{cases} x_a(M) = -2\pi R((1-\beta)M - [(1-\beta)M]) \\ y_a(M) = H\beta M \end{cases} \quad \text{(VII.6)}$$

where M is node number of phyllotaxis lattice, R is cylinder radius, H is pitch of the main genetic helix, and [ ] is the integer part of number.

The distance between two nodes with numbers M and K can be found by the Pythagorean theorem:

$$Q^2 = (x(M) - x(K))^2 + (y(M) - y(K))^2$$

Due to the isomorphism of the cylindrical lattice, we can take K equals zero. Therefore, the distance from the node M along the main genetic helix to the zero node will be:

$$Q^2 = (2\pi R)^2 (\beta M - [\beta M])^2 + (H\beta M)^2 \qquad (VII.7)$$

and along ancillary:

$$Q^2 = (2\pi R)^2 ((1-\beta)M - [(1-\beta)M])^2 + (H\beta M)^2 \qquad (VII.8)$$

From (VII.7) and (VII.8), it follows that Q will be minimum at

$$\begin{cases} M \to \min \\ \beta M \approx [\beta M] \\ (1-\beta)M \approx [(1-\beta)M] \end{cases} \qquad (VII.9)$$

If we add the second and third equations, we get:

$$M = [\beta M] + [(1-\beta)M], \qquad (VII.10)$$

From (VII.9) and (VII.10) we obtain the proportion:

$$\beta M : (1-\beta)M : M \approx [\beta M] : [(1-\beta)M] : ([\beta M] + [(1-\beta)M])$$

or

$$\beta : (1-\beta) : 1 \approx [\beta M] : [(1-\beta)M] : ([\beta M] + [(1-\beta)M]) \qquad (VII.11)$$

As noted in Section II.3, there is a nonlinear transformation of space in which the planar Archimedean spiral can be "pulled" onto to any surface, so that each point of the original spiral corresponds to a point on the transformed helix. Video VII.1[1] shows the conversion of the planar DH-Model into the cylindrical DH-Model, and then unrolls it onto a plane.

As it is a planar phyllotaxis, we first consider a lattice of Fibonacci phyllotaxis on a cylinder, which for simplicity will be called *cylindrical phyllotaxis.*

---

[1] The video is located at https://youtu.be/8ENAjZkjQPs

VII.1. THE CYLINDRICAL FIBONACCI PHYLLOTAXIS

As it is in the planar DH-Model, the cylindrical Fibonacci phyllotaxis will have a divergence coefficient $\beta = \dfrac{1}{\tau^2}$ and accordingly $1 - \beta = \dfrac{1}{\tau}$. Then the lattice nodes of phyllotaxis (VII.5) and (VII.6) will have the coordinates in terms of a two-dimensional Cartesian coordinate system:

$$\begin{cases} x_m(M) = 2\pi R\left(\dfrac{M}{\tau^2} - \left[\dfrac{M}{\tau^2}\right]\right) \\ y_m(M) = H\dfrac{M}{\tau^2} \end{cases} \quad (\text{VII.12})$$

$$\begin{cases} x_a(M) = -2\pi R\left(\dfrac{M}{\tau} - \left[\dfrac{M}{\tau}\right]\right) \\ y_a(M) = H\dfrac{M}{\tau^2} \end{cases}$$

The proportion (VII.11), expressing the minimum distance from the node M to the zero node, will look like this:

$$\left(\left[\dfrac{1}{\tau^2}M\right] + \left[\dfrac{1}{\tau}M\right]\right) : \left[\dfrac{1}{\tau}M\right] : \left[\dfrac{1}{\tau^2}M\right] \approx 1 : \dfrac{1}{\tau} : \dfrac{1}{\tau^2};$$

this formula repeats (V.17):

$$(A + B) : B : A \approx \tau^2 : \tau : 1$$

That is, on the right are three members of a certain integer recurrence series. Reasoning as in Section V.2, we obtain:

$$\begin{cases} \left[\dfrac{1}{\tau^2}M\right] = F_{n-2} \\ \left[\dfrac{1}{\tau}M\right] = F_{n-1} \\ \left[\dfrac{1}{\tau^2}M\right] + \left[\dfrac{1}{\tau}M\right] = F_n \end{cases} \quad (\text{VII.13})$$

From (VII.13), it follows that $M = F_n$, i.e. nodes with numbers s and $s + F_n$ will be nearby.

Arguing similarly to V.3, we find that the nodes of the cylindrical phyllotaxis with numbers $s + i \cdot F_n$ belong to one thread with the degree n.

Let's go back to calculating the distance between two nearby nodes and substitute $\beta = \dfrac{1}{\tau^2}$ and $M = F_n$ into (VII.7):

$$Q^2 = (2\pi R)^2 \left(\dfrac{F_n}{\tau^2} - \left[\dfrac{F_n}{\tau^2} \right] \right)^2 + \left(H \dfrac{F_n}{\tau^2} \right)^2$$

Let's use Binet's formula (II.4):

$$\dfrac{F_n}{\tau^2} = \dfrac{1}{\tau^2} \dfrac{\tau^{n+1} - (-1)^{n+1} \tau^{-(n+1)}}{\sqrt{5}} = \dfrac{\tau^{n-1} - (-1)^{n+1} \tau^{-(n+3)} + (-1)^{n-1} \tau^{-(n-1)} - (-1)^{n-1} \tau^{-(n-1)}}{\sqrt{5}} =$$

$$= \dfrac{\tau^{n-1} - (-1)^{n-1} \tau^{-(n-1)}}{\sqrt{5}} + \dfrac{(-1)^{n-1} \tau^{-(n+1)} (\tau^2 - \tau^{-2})}{\sqrt{5}} = F_{n-2} + (-1)^{n-1} \tau^{-(n+1)} \quad \text{(VII.14)}$$

Because $\tau^{-1} < 1$ and $n > 0$:

$$\left[\dfrac{F_n}{\tau^2} \right] = \left[F_{n-2} + (-1)^{n-1} \tau^{-(n+1)} \right] = F_{n-2} \quad \text{(VII.15)}$$

Substitute (VII.14) and (VII.15) into (VII.8):

$$Q^2 = (2\pi R)^2 (F_{n-2} + (-1)^{n-1} \tau^{-(n+1)} - F_{n-2})^2 + \left(\dfrac{F_n}{\tau^2} H \right)^2 = (2\pi R)^2 \tau^{-2(n+1)} + \left(\dfrac{F_n}{\tau^2} H \right)^2$$

(VII.16)

Let's divide both sides of the equality (VII.16) by the square of the pitch of the main genetic spiral H^2:

$$\dfrac{Q^2}{H^2} = \left(2\pi \dfrac{R}{H} \right)^2 \tau^{-2(n+1)} + \left(\dfrac{F_n}{\tau^2} \right)^2$$

Let's use approximate Binet's formula $F_n \approx \dfrac{\tau^{n+1}}{\sqrt{5}}$:

$$\dfrac{Q}{H} = \sqrt{ \left(\dfrac{2\pi}{\sqrt{5} F_n} \dfrac{R}{H} \right)^2 + \left(\dfrac{F_n}{\tau^2} \right)^2 } \quad \text{(VII.17)}$$

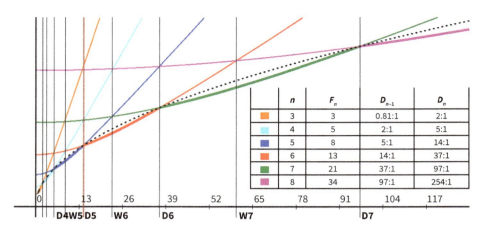

Figure VII.2. The graph of the distance between nearby nodes for families of threads as the function of the RH-ratio.

Let's compare the last formula with (V.22): the role of the node number in (VI.9) performs the ratio of the radius of the cylinder to the pitch of the helix; therefore, for brevity, we will call *the RH-ratio*. The value $2\pi R / H$ is equal to the cotangent of the helix angle (Figure VII.1).

In the same way, as to the planar DH-Model, let's construct, according to (VII.17), a family of graphs (Figure VII.2) of the dependence of the distance between nearby nodes on the RH-ratio for the families of threads. Each thread family has its own color curve.

Similarly, to the planar DH-Model, let's find the boundaries of the ranges of best visibility as the points of intersection of the graphs of the dependence of the distance between the nearby nodes on the RH-ratio for threads of the n-th and the $(n + 1)$-th degrees:

$$\left(\frac{2\pi}{\sqrt{5}F_n}D_n\right)^2 + \left(\frac{F_n}{\tau^2}\right)^2 = \left(\frac{2\pi}{\sqrt{5}F_{n+1}}D_n\right)^2 + \left(\frac{F_{n+1}}{\tau^2}\right)^2$$

$$D_n = \frac{\sqrt{5}F_{n+1}F_n}{2\pi\tau^2} \approx \frac{\sqrt{5}F_n^2}{2\pi\tau}, \qquad \text{(VII.18)}$$

where D_n is the RH-ratio (Table VII.1), at which the distance between nearby nodes will be the same for threads of the n-th and the $(n + 1)$-th degrees.

Similarly, to the planar DH-Model, let's find the values of the RH-ratio at which the visible pair of parastichies changes, that is called "the phyllotaxis rises." This occurs when the distance between the nodes of the phyllotactic lattice is the same for threads of the $(n - 1)$-th and the $(n + 1)$-th degrees. In Figure VII.2, these are the intersection points of the graphs corresponding to the $(n-1)$-th and the $(n+1)$-th family of threads:

$$\left(\frac{2\pi}{\sqrt{5}F_{n-1}}W_n\right)^2 + \left(\frac{F_{n-1}}{\tau^2}\right)^2 = \left(\frac{2\pi}{\sqrt{5}F_{n+1}}W_n\right)^2 + \left(\frac{F_{n+1}}{\tau^2}\right)^2$$

$$W_n = \frac{\sqrt{5}F_{n+1}F_{n-1}}{2\pi\tau^2} \approx \frac{\sqrt{5}F_n^2}{2\pi\tau^2}, \tag{VII.19}$$

where W_n is the RH-ratio (Table VII.2), at which the distance between nearby nodes will be the same for threads of the $(n-1)$-th and the $(n+1)$-th degrees.

Table VII.1. Diapasons of the cylindrical Fibonacci phyllotaxis.

The number of diapasons N	The number of threads in the n-th family F_n	Color of the n-th threads family according to Figure VII.2	The lower border of the n-th diapason D_{n-1}	The upper border of the n-th diapason D_n
3	3	Orange	2:1	5.4:1
4	5	Cyan	5.4:1	14:1
5	8	Blue	14:1	37:1
6	13	Red	37:1	97:1
7	21	Green	97:1	254:1
8	34	Magenta	254:1	665:1

Table VII.2. "Rise" of the cylindrical Fibonacci phyllotaxis.

	The RH-ratio, in which "the phyllotaxis rises" occurs	Changing the index of parastichies
n	W_n	$(F_{n-1}, F_n) \rightarrow (F_n, F_{n+1})$
4	3.34 : 1	(3, 5) → (5, 8)
5	8.7 : 1	(5, 8) → (8, 13)
6	23 : 1	(8, 13) → (13, 21)
7	60 : 1	(13, 21) → (21, 34)
8	157 : 1	(21, 34) → (34, 55)

From (VII.17), it follows that with an increase (decrease) in the RH-ratio by more than τ^2 ($\approx 2{,}1618$) times, the visual effect of an increase (decrease) in the index of visible parastichies will necessarily occur.

Similarly, to Section V.6, let's suppose primordia are like incompressible, but a non-deformable silicone discs which center coincides with its corresponding node, and these disks fill the space of the phyllotactic lattice to its maximum. Let us find the approximate function of the dependence of the radius of such a disk on the RH-ratio. The graph of this function passes through the border points of each diapason (VII.18) (dashed line in Figure VII.2):

$$d(H,R) = \begin{cases} \dfrac{R}{H} = \dfrac{\sqrt{5}}{2\pi\tau}F_n^2 \\ \sqrt{\left(\dfrac{2\pi}{\sqrt{5}F_n}R\right)^2 + \left(\dfrac{F_n}{\tau^2}H\right)^2} \end{cases} \quad \text{(VII.20)}$$

Let's exclude n, by substituting $F_n^2 = \dfrac{2\pi\tau}{\sqrt{5}}\dfrac{R}{H}$, then:

$$dr(H,R) = \sqrt{\dfrac{\sqrt{5}}{2\pi\tau}\dfrac{H}{R}\left(\dfrac{2\pi}{\sqrt{5}}R\right)^2 + \dfrac{2\pi\tau}{\sqrt{5}}\dfrac{R}{H}\left(\dfrac{H}{\tau^2}\right)^2} = \sqrt{\dfrac{2\pi}{\tau^2}RH} \quad \text{(VII.21)}$$

Similarly, to Section V.7, let's find the angle between the tangents to the oncoming parastichies of the n-th and the $(n+1)$-th families. On the cylindrical DH-Model, this is much simpler, since the parastichies on the unrolled cylinder are visually perceived as straight. That is, the parastichies coincide with the tangent lines.

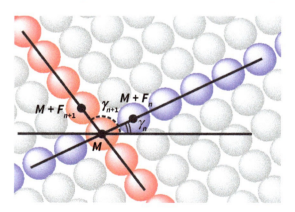

Figure VII.3. The angle between the parastichies of the n-th and the $(n+1)$-th families.

The angle between the parastichy of the n-th family and the ordinate axis can be found through the cotangent function:

$$ctan(\gamma_n) = \dfrac{x_{M+F_n} - x_M}{y_{M+F_n} - y_M}$$

Due to the isotropy of the cylindrical DH-Model, it is possible to take the node M as zero and use (VII.12):

$$ctan(\gamma_n) = \frac{x_{F_n} - 0}{y_{F_n} - 0} = \frac{2\pi R\left(\frac{F_n}{\tau^2} - \left[\frac{F_n}{\tau^2}\right]\right)}{H\frac{F_n}{\tau^2}} = \frac{2\pi R(-1)^{n-1}\tau^{-(n+1)}}{H\frac{F_n}{\tau^2}} = (-1)^{n-1}\frac{2\pi\tau^2}{\tau^{n-1}F_n}\frac{R}{H}$$

For the parastichy of the $(n+1)$-th family:

$$ctan(\gamma_{n+1}) = \frac{x_{F_{n+1}} - 0}{y_{F_{n+1}} - 0} = \frac{2\pi R\left(\frac{F_{n+1}}{\tau^2} - \left[\frac{F_{n+1}}{\tau^2}\right]\right)}{H\frac{F_{n+1}}{\tau^2}} = \frac{2\pi R(-1)^n \tau^{-(n+2)}}{H\frac{F_{n+1}}{\tau^2}} = (-1)^n \frac{2\pi}{\tau^n F_{n+1}}\frac{R}{H}$$

Figure VII.3 shows that the angle between the parastichies of the n-th and the $(n+1)$-th families will be equal to the difference of the angles γ_n and γ_{n+1}:

$$\Delta\gamma = |\gamma_{n+1} - \gamma_n| = \left|arcctan\left((-1)^{n-1}\frac{2\pi}{\tau^{n-1}F_n}\frac{R}{H}\right) - arcctan\left((-1)^n \frac{2\pi}{\tau^n F_{n+1}}\frac{R}{H}\right)\right| =$$

$$= arcctan\left(\frac{2\pi}{\tau^{n-1}F_n}\frac{R}{H}\right) + arcctan\left(\frac{2\pi}{\tau^n F_{n+1}}\frac{R}{H}\right) \qquad (VII.22)$$

From the last formula, it follows that the angle between the oncoming parastichies depends on the RH-ratio only (Figure VII.4).

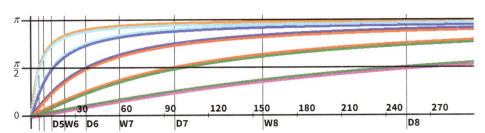

Figure VII.4. The graph of the angle between the oncoming parastichies as the function of the RH-ratio.

Indeed, if in (VII.22), we substitute the boundaries of the ranges from (VII.18), then we obtain:

$$\Delta\gamma = \approx arcctg(\tau) + arcctg\left(\frac{1}{\tau}\right) = \frac{\pi}{2}$$

The phyllotactic lattice on Archimedean helices has structural recursiveness and structural self-similarity due to the lattice isotropy.

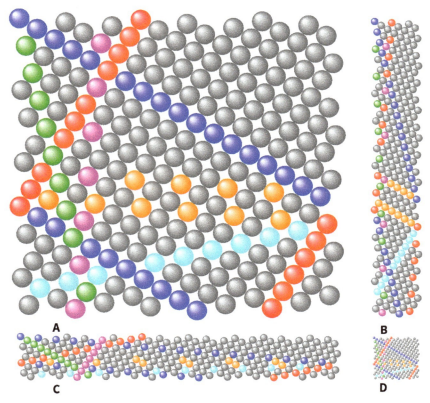

Figure VII.5. The phyllotactic lattice (225 nodes) on a planar representation of the cylindrical phyllotaxis with different values of the RH-ratio: A – 13:1, B – 2:1, C – 102:1, D – 13:1.

Figure VII.5.A shows that the planar representation of the cylindrical phyllotaxis of the phyllotactic lattice with the RH-ratio 13:1. The color coding of phyllotactic threads corresponds to Table VII.1.1.

The planar representation of the cylindrical phyllotaxis of the lattice in Figure VII.5.B was obtained by horizontal compression of the lattice in Figure VII.5.A by 6.8 times. Similarly, the pattern in Figure VII.5.C was obtained by vertical compression of the lattice in Figure VII.5.A by 6.8 times; the phyllotactic pattern in Figure VII.5.D was obtained by compressing Figure VII.5.B by 6.8 times vertically or by compressing Figure VII.5.C by 6.8 times horizontally.

Video VII.2[2] shows, that when stretching or compressing horizontally or vertically the same phyllotactic lattice, the nodes "fold" into parastichies and

---

[2] The video is located at https://youtu.be/6ofqTVtP4EE

"disintegrate". It is important that the structure of the phyllotactic lattice remains constant. As it is in the case of the planar DH-Model, the right- and left-handed sloping straight lines are only the result of human perception, which combines nearby lattice nodes into "parastichies".

In this video, it is clearly seen that no abrupt transition occurs with "the phyllotaxis rises". This refutes the stereotype existing among some researchers about a qualitative change in the lattice with "the phyllotaxis rises".

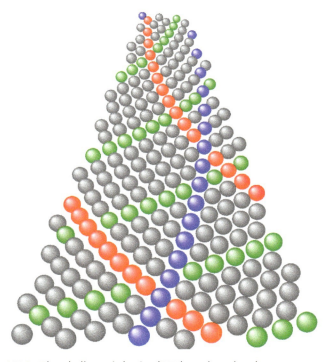

Figure VII.6. The phyllotactic lattice (225 knots) on the planar representation of the cone phyllotaxis.

Many plants, such as fir, have a trunk whose diameter decreases from bottom to top. Therefore, the upper part of the trunk may have a different index of visible parastichy than the lower. This visual effect is clearly visible on the phyllotactic lattice on a cone. When the lattice of the cone phyllotaxis unrolled onto a plane (Figure VII.6), then in the lower part of the cone, the red and blue threads are the visible opposed parastichy pair, and in the upper part - the red and green threads. Video VII.3[3] shows

---

[3] The video is located at https://youtu.be/S-EFLHDolUc

the transition from one visible opposed parastichy pair to another during the transition from the phyllotactic lattice on the cylinder to the lattice on the cone.

VII.2. THE CYLINDRICAL PHYLLOTAXIS GENERATED BY A NON-MULTIPLE RECURRENT SEQUENCE

Let's consider the planar representation of the cylindrical phyllotaxis generated by a non-multiple recurrent sequence:

$$G_1, G_2, G_2 + G_1, 2G_2 + G_1, 3G_2 + 2G_1, 5G_2 + 3G_1, \ldots ,$$

where $1 < G_1$, $2G_1 \leq G_2$; $\gcd(G_1, G_2) = 1$, $G_1, G_2 \in N$. Then the nodes of this phyllotactic lattice will have the coordinates in terms of a two-dimensional Cartesian coordinate system:

$$\begin{cases} x(M) = 2\pi R(\beta M - [\beta M]) \\ y(M) = H\beta M \end{cases}$$

Such a phyllotactic lattice will have G_n threads (parastichies) in every n-th family. Then the neighboring nodes in each thread will be nodes with numbers M and $M + G_n$, and the distance between them:

$$Q^2 = (2\pi R)^2 (\beta G_n - [\beta G_n])^2 + (H\beta G_n)^2 \tag{VII.23}$$

From (VII.23), it is follows that for constant R and H, Q will be minimal for:

$$\beta G_n \approx [\beta G_n]$$

Let's use the formulas (II.10)–(II.14) from Section on the expansion of the recurrence series in the sum:

$$G_n = A_n + B_n$$

where A_n and B_n are terms of some integer recurrence series, moreover, $A_n < B_n$ because $0 < \beta < 0.5$. Define A_n as

$$A_n = [\beta G_n] \approx \beta G_n.$$

Then:

$$\beta = \lim_{n \to \infty} \frac{A_n}{G_n} = \frac{A_2 + \dfrac{A_1}{\tau}}{G_2 + \dfrac{G_1}{\tau}} \tag{VII.24}$$

Let's use Binet's formula (II.7) and (VII.24) to calculate $|[\beta G_n] - \beta G_n|$:

$$|[\beta G_n] - \beta G_n| = |A_n - \beta G_n| =$$

$$= \left| \left(\frac{A_2 + A_1 \cdot \tau^{-1}}{\sqrt{5}} \tau^{n-1} - \frac{A_2 - A_1 \cdot \tau}{\sqrt{5}} (-\tau)^{-n+1} \right) - \right.$$

$$\left. - \frac{A_2 + \dfrac{A_1}{\tau}}{G_2 + \dfrac{G_1}{\tau}} \left(\frac{G_2 + G_1 \cdot \tau^{-1}}{\sqrt{5}} \tau^{n-1} - \frac{G_2 - G_1 \cdot \tau}{\sqrt{5}} (-\tau)^{-n+1} \right) \right| =$$

$$= \frac{1}{\sqrt{5}} \frac{\tau^{-n+1}}{G_2 + \dfrac{G_1}{\tau}} \left| \left(G_2 + \frac{G_1}{\tau} \right)(A_2 - A_1 \cdot \tau) - \left(A_2 + \frac{A_1}{\tau} \right)(G_2 - G_1 \cdot \tau) \right| \approx$$

$$\approx \frac{1}{\sqrt{5} G_n} |A_1 G_2 - A_2 G_1| \qquad (VII.25)$$

As it is in planar phyllotaxis, $|[\beta G_n] - \beta G_n|$ will be minimum at

$$|A_1 G_2 - A_2 G_1| = 1 \qquad (VII.26)$$

Substitute (VII.25) and (VII.26) into (VII.23):

$$Q^2 = (2\pi R)^2 \left(\frac{1}{\sqrt{5} G_n} \right)^2 + (H \beta G_n)^2 \qquad (VII.27)$$

Let's divide both sides of equality (VII.27) by the square of the pitch of the main genetic helix H^2 and extract the root from both sides of the equation:

$$\frac{Q}{H} = \sqrt{(\beta G_n)^2 + \left(\frac{2\pi}{\sqrt{5} G_n} \right)^2 \left(\frac{R}{H} \right)^2}$$

As well as in a cylindrical Fibonacci phyllotaxis, the RH-ratio performs the role of the node number in the cylindrical phyllotaxis generated by a non-multiple recurrent sequence. By analogy with (VII.18)–(VII.1.19), let's find the boundaries of the diapasons for cylindrical phyllotaxis generated by a non-multiple a recurrent sequence:

$$D_n = \frac{\sqrt{5} \beta G_{n+1} G_n}{2\pi} \approx \frac{\sqrt{5} \tau \beta G_n^2}{2\pi}$$

The values of the RH-ratio where "the phyllotaxis rises" occurs:

$$W_n = \frac{\sqrt{5}\beta G_{n+1} G_{n-1}}{2\pi} \approx \frac{\sqrt{5}\beta G_n^2}{2\pi}$$

The value of the diameter of the element from the RH-ratio:

$$dr(H,R) = \sqrt{\frac{\sqrt{5}\tau\beta}{2\pi}\frac{H}{R}\left(\frac{2\pi}{\sqrt{5}}R\right)^2 + \frac{2\pi}{\sqrt{5}\tau\beta}\frac{R}{H}(\beta H)^2} = \sqrt{2\pi\beta R H}$$

The value of the angle between the opposite parastichies from the RH-ratio:

$$\Delta\gamma \approx \text{arcctan}\left(\frac{2\pi}{\tau^{n-1}G_n}\frac{R}{H}\right) + \text{arcctan}\left(\frac{2\pi}{\tau^n G_{n+1}}\frac{R}{H}\right)$$

If in the formulas of this Section we take $G_1 = 1$ and $G_2 = 2$, then we get the formulas from Section VII.1. This allows us to conclude that cylindrical phyllotactic lattices generated by the Fibonacci sequence are a particular case of phyllotactic lattices generated by non-multiple recurrent sequence.

These videos show the process of compression and extension of cylindrical phyllotactic lattices generated by non-multiple recurrent sequence.

G_1	G_2	Video	URL
1	3	Video VII.4	https://youtu.be/iEgZjy-xLMk
1	4	Video VII.5	https://youtu.be/e7XxMtTnEaM
1	5	Video VII.6	https://youtu.be/ChyX26h2QuI
2	5	Video VII.7	https://youtu.be/z9-jxViiy20
2	7	Video VII.8	https://youtu.be/nsEJTaiUaFM
3	7	Video VII.9	https://youtu.be/K-rssHxtzoY

VII.3. THE CYLINDRICAL PHYLLOTAXIS GENERATED BY THE GENERALIZED RECURRENT SEQUENCE

Let's consider the planar representation of the cylindrical phyllotaxis generated by the generalized recurrent sequence:

$$jG_1, jG_2, jG_2 + jG_1, 2jG_2 + jG_1, 3jG_2 + 2jG_1, 5jG_2 + 3jG_1, \ldots$$

where $1 \leq G_1$, $2G_1 \leq G_2$, $1 \leq j$, $\gcd(G_2, G_1) = 1$, $G_1, G_2, j \in \mathbb{N}$.

By analogy with the planar DH-Model, the lattice of such phyllotaxis will be formed by j pairs of genetic helices:

$$\begin{cases} x_{m,k}(\theta) = R \cdot \cos\left(2\pi\left(\frac{\beta}{j}\theta + \frac{k}{j}\right)\right) \\ y_{m,k}(\theta) = R \cdot \sin\left(2\pi\left(\frac{\beta}{j}\theta + \frac{k}{j}\right)\right) \\ z_{m,k}(\theta) = H_m \cdot \beta\theta \end{cases} \quad \text{(VII.28)}$$

$$\begin{cases} x_{a,k}(\theta) = R\cos\left(2\pi\left(\left(1-\frac{\beta}{j}\right)\theta + \frac{k}{j}\right)\right) \\ y_{a,k}(\theta) = -R\sin\left(2\pi\left(\left(1-\frac{\beta}{j}\right)\theta + \frac{k}{j}\right)\right) \\ z_{a,k}(\theta) = H_a \cdot (1-\beta)\theta \end{cases} \quad \text{(VII.29)}$$

where $0 \le k < j, k \in N$, and divergence coefficient $\beta = \dfrac{A_2 + \dfrac{A_1}{\tau}}{G_2 + \dfrac{G_1}{\tau}}$.

Then the nodes of this planar representation of the cylindrical phyllotaxis will have coordinates in terms of a two-dimensional Cartesian coordinate system:

$$\begin{cases} x_k(M) = 2\pi R\left(\frac{1}{j}\beta M - \frac{1}{j}[\beta M] + \frac{k}{j}\right) \\ y_k(M) = H\beta M \end{cases} \quad \text{(VII.30)}$$

As in planar multi-pair phyllotaxis, in multi-pair cylindrical phyllotaxis (VII.30), there is no continuous numbering of nodes. The nodes generated by each pair of genetic helices have their own numbering. Therefore, when we say the node M, we will mean one of the nodes with the number M.

Figure VII.7 shows the structure of the three-pair cylindrical phyllotactic lattice which is unrolled on the plane. The lattice contains three pairs of genetic helices (VII.28)–(VII.29). Each of the pairs of genetic helices forms nodes with the divergence angle $2\pi\dfrac{\beta}{j} = 2\pi\dfrac{1}{j}\dfrac{A_2 + \dfrac{A_1}{\tau}}{G_2 + \dfrac{G_1}{\tau}}$. Figure VII.7.A shows the nodes formed by the first pair of genetic helices, and Figure VII.7.B and Figure VII.7.C show the nodes of the other two pairs, respectively. Figure VII.7.D depicts all the nodes generated by three pairs of the genetic helices.

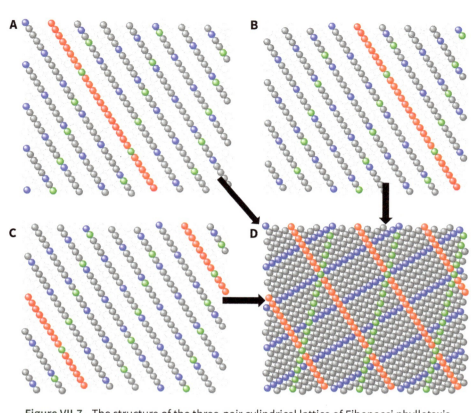

Figure VII.7. The structure of the three-pair cylindrical lattice of Fibonacci phyllotaxis unrolled on a plane.

Such a phyllotactic lattice has jG_n threads (parastichies) in each n-th family, but the neighboring nodes in each thread will be the nodes M and $M + G_n$, and the distance between them will be

$$Q^2 = (2\pi R)^2 \left(\frac{1}{j}\beta M - \frac{1}{j}[\beta M] \right)^2 + (H\beta G_n)^2 \qquad \text{(VII.31)}$$

Similarly to the previous Section, from (VII.30), it is follows that for constant R and H, Q will be minimal for:

$$\beta G_n \approx [\beta G_n]$$

Let's use the formulas (II.10)–(II.14) from the Section on the expansion of the recurrence series in the sum:

$$G_n = A_n + B_n$$

where A_n and B_n are terms of some integer recurrence series, moreover, $A_n < B_n$ because $0 < \beta < 0.5$. Define A_n as:

$$A_n = [\beta G_n] \approx \beta G_n$$

Then:

$$\beta = \lim_{n \to \infty} \frac{A_n}{G_n} = \frac{A_2 + \dfrac{A_1}{\tau}}{G_2 + \dfrac{G_1}{\tau}}$$

Similarly to (VII.25):

$$\left| \frac{1}{j} \beta M - \frac{1}{j}[\beta M] \right| \approx \frac{1}{\sqrt{5} j G_n} |A_1 G_2 - A_2 G_1| \tag{VII.32}$$

As it is in planar phyllotaxis, $\left| \left[\dfrac{\beta}{j} G_n\right] - \dfrac{\beta}{j} G_n \right|$ will be minimum at

$$|A_1 G_2 - A_2 G_1| = 1 \tag{VII.33}$$

Substitute (VII.32) and (VII.33) into (VII.31):

$$Q^2 = (2\pi R)^2 \left(\frac{1}{\sqrt{5} j G_n} \right)^2 + (H \beta G_n)^2 \tag{VII.34}$$

Let's divide both sides of equality (VII.34) by the square of the pitch of the main genetic spiral H^2 and extract the root from both sides of the equation:

$$\frac{Q}{H} = \sqrt{(\beta G_n)^2 + \left(\frac{2\pi}{\sqrt{5} j G_n} \right)^2 \left(\frac{R}{H} \right)^2 }$$

As well as it is in a cylindrical Fibonacci phyllotaxis, the RH-ratio performs the role of the node number in the cylindrical phyllotaxis generated by a generalized recurrent sequence. By analogy with (VII.18)–(VII.1.19), let's find the boundaries of the ranges for cylindrical phyllotaxis generated by a recurrent sequence.

The boundaries of the diapasons:

$$D_n = \frac{\sqrt{5} j \beta G_{n+1} G_n}{2\pi} \approx \frac{\sqrt{5} j \tau \beta G_n^2}{2\pi}$$

Nodes of "the phyllotaxis rises":

$$W_n = \frac{\sqrt{5}j\beta G_{n+1}G_{n-1}}{2\pi} \approx \frac{\sqrt{5}j\beta G_n^2}{2\pi}$$

The diameter of the element of the phyllotactic lattice:

$$dr(H,R) = \sqrt{\frac{\sqrt{5}j\tau\beta\,H}{2\pi\,R}\left(\frac{2\pi}{\sqrt{5}j}R\right)^2 + \frac{2\pi}{\sqrt{5}j\tau\beta\,H}\frac{R}{H}(\beta H)^2} = \sqrt{\frac{2\pi\beta RH}{j}}$$

The angle between the opposed parastichies:

$$\gamma \approx arcctg\left(\frac{2\pi\tau^2}{\sqrt{5}jG_{n+1}^2}\frac{R}{H}\right) + arcctg\left(\frac{2\pi\tau^2}{\sqrt{5}jG_n^2}\frac{R}{H}\right)$$

If in the formulas of this Section, let's take $j = 1$, then we get the formulas from Section VII.2. This allows us to conclude that cylindrical phyllotactic lattices generated by the non-multiple recurrent sequence are a particular case of phyllotactic lattices generated by generalized recurrent sequence.

These videos show the process of compression and extension of cylindrical phyllotactic lattices generated by generalized recurrent sequence.

j	G_1	G_2	Figure	URL
2	1	2	Video VII.10	https://youtu.be/7jqLGss_RY4
2	1	3	Video VII.11	https://youtu.be/EtQzSMmoLC0
3	1	2	Video VII.12	https://youtu.be/PtbcBof-Gps

VIII

PHYLLOTACTIC LATTICES WITH RATIONAL DIVERGENCE COEFFICIENT

In the previous Sections, lattices of DH-Model generated by recurrent rows were analyzed. A characteristic feature of these lattices was that the divergence coefficient was an irrational number. Let's consider which lattices appear in the DH-Model for some rational values of the divergence coefficient. Let me remind you that in the description of the DH-Model the boundaries of the divergence coefficient were defined as $0 \le \beta \le 0.5$.

For simplicity, let's do an analysis of lattices on Archimedean spirals.

VIII.1. PHYLLOTACTIC LATTICES WITH THE DIVERGENCE COEFFICIENT $\beta = 0$

At $\beta = 0$, the genetic helices (V.1) and (V.2) for a planar phyllotaxis lattice will be:

$$\begin{cases} x_m(\theta) = \theta \\ y_m(\theta) = 0 \end{cases} \qquad (\text{VIII.1})$$

$$\begin{cases} x_a(\theta) = \theta \cdot \cos(2\pi\theta) \\ y_a(\theta) = -\theta \cdot \sin(2\pi\theta) \end{cases} \qquad (\text{VIII.2})$$

As it follows from (VIII.1), the main genetic spiral degenerated into a ray coinciding with the positive axis OX, and the ancillary genetic spiral (VIII.2) becomes a classical the Archimedean spiral twisted clockwise (Figure VIII.1.A). The nodes of such a lattice have coordinates $(M, 0)$ in a Cartesian coordinate system, where $M \in \mathbb{N}$.

The nodes in the cylindrical phyllotactic lattice with a divergence coefficient $\beta = 0$ (Figure VIII.1.B) will have coordinates $(R, 0, H \cdot M)$, where R is the radius of the cylinder, H is the step of the ancillary genetic helix, and M is the node number.

It should be noted that phyllotactic patterns corresponding to this lattice do not occur in nature; therefore, this lattice is only a theoretical assumption.

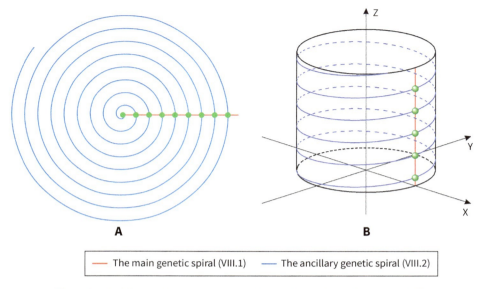

Figure VIII.1. The phyllotactic lattice at $\beta = 0$: planar (**A**) and cylindrical (**B**).

VIII.2. PHYLLOTACTIC LATTICES WITH THE DIVERGENCE COEFFICIENT $\beta = 0.5$

At $\beta = 0.5$, the genetic helices (V.1) and (V.2) for the planar phyllotactic lattice will be:

$$\begin{cases} x_m(\theta) = \theta \cdot \cos(\pi\theta) \\ y_m(\theta) = \theta \cdot \sin(\pi\theta) \end{cases} \quad \text{(VIII.3)}$$

$$\begin{cases} x_a(\theta) = \theta \cdot \cos(\pi\theta) \\ y_a(\theta) = -\theta \cdot \sin(\pi\theta) \end{cases} \quad \text{(VIII.4)}$$

From (VIII.3) and (VIII.4), it follows that the main and ancillary genetic spirals are reflection symmetry relatively to the OX axis and have the same pitch. The nodes of such a lattice have coordinates $((-1)^M M, 0)$ in a Cartesian coordinate system, where $M \in \mathbb{N}$.

The nodes in the cylindrical phyllotactic lattice with a divergence coefficient $\beta = 0.5$ (Figure VIII.2.B) will have the coordinates $\left((-1)^M R, 0, \dfrac{H}{2} M\right)$, where R is the radius of the cylinder, H is the step of the ancillary genetic helix, and M is the node number.

If we compare Figure VIII.2.B to Figure III.4.A, we can verify that the lattice at $\beta = 0.5$ corresponds to the distichous or an alternate phyllotactic pattern.

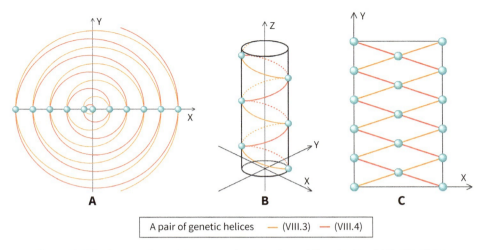

Figure VIII.2. The phyllotactic lattice at β = 0.5: planar (**A**), cylindrical (**B**), the planar representation of the cylindrical phyllotaxis (**C**).

VIII.3. TWO-PAIR PHYLLOTACTIC LATTICES WITH THE DIVERGENCE COEFFICIENT $\beta = 0.5$

Let me recall that the lattice of the two-pair phyllotaxis consists of two pairs of genetic spirals rotated to each other by π. Therefore, to the pair of genetic spirals (VIII.3) and (VIII.4), a couple of genetic spirals are added rotated by π:

$$\begin{cases} x_{2,m}(\theta) = \theta \cdot \cos(\pi\theta + \pi) = -\theta \cdot \cos(\pi\theta) \\ y_{2,m}(\theta) = \theta \cdot \sin(\pi\theta + \pi) = -\theta \cdot \sin(\pi\theta) \end{cases} \quad \text{(VIII.5)}$$

$$\begin{cases} x_{2,a}(\theta) = \theta \cdot \cos(\pi\theta + \pi) = -\theta \cdot \cos(\pi\theta) \\ y_{2,a}(\theta) = -\theta \cdot \sin(\pi\theta + \pi) = \theta \cdot \sin(\pi\theta) \end{cases} \quad \text{(VIII.6)}$$

Such a lattice will have twice as many nodes and each number M will correspond to two nodes that have coordinates in a Cartesian coordinate system ($\pm M$, 0), where $M \in N$.

The nodes in the cylindrical phyllotactic lattice with $\beta = 0.5$ and $j = 2$ (Figure VIII.3.B) will have coordinates $\left(\pm R, 0, \dfrac{H}{2}M\right)$, where R is the radius of the cylinder, H is the step of the genetic helices, and M is the node number.

If we compare Figure VIII.3.B and Figure III.4.B, we can verify that the lattice at $\beta = 0.5$ and $j = 2$ corresponds to the opposite superposed phyllotaxis.

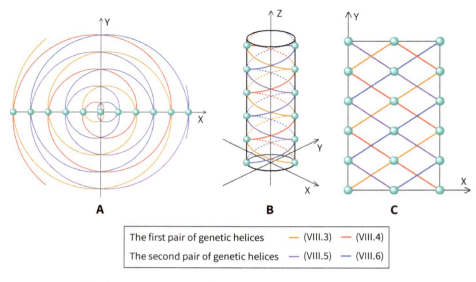

| The first pair of genetic helices | — (VIII.3) | — (VIII.4) |
| The second pair of genetic helices | — (VIII.5) | — (VIII.6) |

Figure VIII.3. The phyllotactic lattice at $\beta = 0.5$ and $j = 2$: planar (**A**), cylindrical (**B**), the planar representation of the cylindrical phyllotaxis (**C**).

VIII.4. MULTI-PAIR PHYLLOTACTIC LATTICES WITH THE DIVERGENCE COEFFICIENT $\beta = 0.5$

The planar phyllotactic lattice with the divergence coefficient $\beta = 0.5$ and $j = 3$ is generated by three pairs of genetic helices rotated by $\dfrac{2\pi}{3}$ to each other:

$$\begin{cases} x_{k,m}(\theta) = \theta \cdot \cos\left(\pi\theta + \dfrac{2\pi}{3}k\right) \\ y_{k,m}(\theta) = \theta \cdot \sin\left(\pi\theta + \dfrac{2\pi}{3}k\right) \end{cases}, \text{ where } k = \overline{0,2},\ k \in N \qquad (\text{VIII.7})$$

$$\begin{cases} x_{k,a}(\theta) = \theta \cdot \cos\left(-\pi\theta + \dfrac{2\pi}{3}k\right) \\ y_{k,a}(\theta) = \theta \cdot \sin\left(-\pi\theta + \dfrac{2\pi}{3}k\right) \end{cases}, \text{ where } k = \overline{0,2},\ k \in N \qquad (\text{VIII.8})$$

Such a lattice will have three times more nodes and each number M will correspond to three nodes that have coordinates in a Cartesian coordinate system $((-1)^M M, 0), \left((-1)^{M-1}\dfrac{1}{2}M,\ \pm\dfrac{\sqrt{3}}{2}M\right)$, where $M \in N$.

When comparing Figure VIII.4.B and Figure III.4.D, we can verify that the lattice at $\beta = 0.5$ and $j = 3$ corresponds to the 3-whorled phyllotaxis.

The reader can independently construct phyllotactic lattices with $\beta = 0.5$ for $j = 4$ or $j = 5$ to see the patterns, which correspond to 4-whorled or 5-whorled phyllotaxis (Figure III.4.E).

Phyllotactic Lattices with Rational Divergence Coefficient

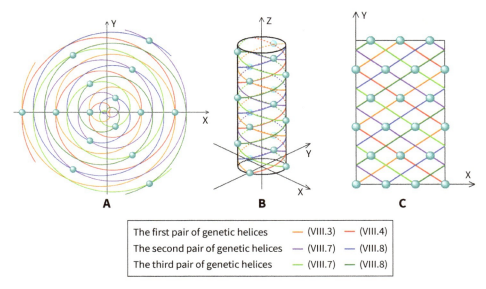

The first pair of genetic helices	— (VIII.3)	— (VIII.4)
The second pair of genetic helices	— (VIII.7)	— (VIII.8)
The third pair of genetic helices	— (VIII.7)	— (VIII.8)

Figure VIII.4. The phyllotactic lattice at $\beta = 0.5$ and $j = 3$: planar (**A**), cylindrical (**B**), the planar representation of the cylindrical phyllotaxis (**C**).

VIII.5. PHYLLOTACTIC LATTICES WITH THE DIVERGENCE COEFFICIENT $\beta = 0.25$

At $\beta = 0.25$, the genetic helices (V.1) and (V.2) for the planar phyllotactic lattice will be:

$$\begin{cases} x_m(\theta) = \theta \cdot \cos\left(\dfrac{\pi}{2}\theta\right) \\ y_m(\theta) = \theta \cdot \sin\left(\dfrac{\pi}{2}\theta\right) \end{cases} \quad \text{(VIII.9)}$$

$$\begin{cases} x_a(\theta) = \theta \cdot \cos\left(-\dfrac{3\pi}{2}\theta\right) \\ y_a(\theta) = \theta \cdot \sin\left(-\dfrac{3\pi}{2}\theta\right) \end{cases} \quad \text{(VIII.10)}$$

From (VIII.9) and (VIII.10), it is follows that in one revolution of the main genetic spiral, the ancillary makes 3 rotations. The nodes of such a lattice have these coordinates in a Cartesian coordinate system:

$$\begin{cases} x(M) = M \cdot \cos\left(\dfrac{\pi}{2}M\right) \\ y(M) = M \cdot \sin\left(\dfrac{\pi}{2}M\right) \end{cases}, \quad M \in N$$

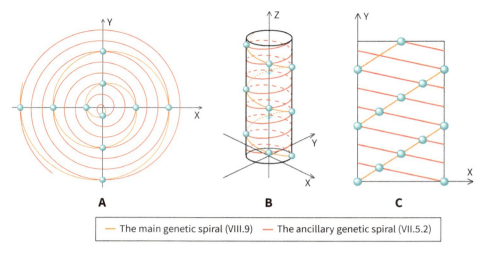

— The main genetic spiral (VIII.9)   — The ancillary genetic spiral (VII.5.2)

Figure VIII.5. The phyllotactic lattice at $\beta = 0.25$: planar (**A**), cylindrical (**B**), the planar representation of the cylindrical phyllotaxis (**C**).

The nodes in the cylindrical phyllotactic lattice with a divergence coefficient $\beta = 0.25$ (Figure VIII.5.B) will have the coordinates:

$$\begin{cases} x(M) = R \cdot \cos\left(\dfrac{\pi}{2} M\right) \\ y(M) = R \cdot \sin\left(\dfrac{\pi}{2} M\right), \\ z(M) = \dfrac{H}{4} M \end{cases}$$

where R is the radius of the cylinder, H is the step of the ancillary genetic helix, and M is the node number.

It should be noted that phyllotactic patterns corresponding to this lattice do not occur in nature, therefore, this lattice is only a theoretical assumption.

VIII.6. THE TWO-PAIR PHYLLOTACTIC LATTICE WITH THE DIVERGENCE COEFFICIENT $\beta = 0.25$

As in the case of the two-pair phyllotaxis with $\beta = 0.5$, to the pair of genetic spirals (VIII.9) and (VIII.10), add a couple genetic spirals, which is rotated by π.

$$\begin{cases} x_{2,m}(\theta) = \theta \cdot \cos\left(\dfrac{\pi}{2}\theta + \pi\right) = -\theta \cdot \cos\left(\dfrac{\pi}{2}\theta + \pi\right) \\ y_{2,m}(\theta) = \theta \cdot \sin\left(\dfrac{\pi}{2}\theta + \pi\right) = -\theta \cdot \sin\left(\dfrac{\pi}{2}\theta + \pi\right) \end{cases} \quad \text{(VIII.11)}$$

$$\begin{cases} x_{2,a}(\theta) = \theta \cdot \cos\left(-\dfrac{\pi}{2}\theta + \pi\right) = \theta \cdot \cos\left(-\dfrac{\pi}{2}\theta + \pi\right) \\ y_{2,a}(\theta) = \theta \cdot \sin\left(-\dfrac{\pi}{2}\theta + \pi\right) = \theta \cdot \sin\left(-\dfrac{\pi}{2}\theta + \pi\right) \end{cases} \quad \text{(VIII.12)}$$

Such a lattice will have twice as many nodes and each number M will correspond to two nodes that have coordinates in a Cartesian coordinate system ($\pm(M + 0.5)$, 0) and ($\pm M$, 0), where $M \in N$.

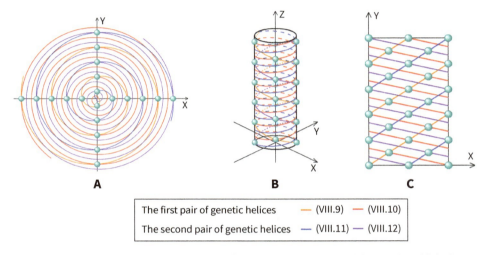

Figure VIII.6. The phyllotactic lattice at $\beta = 0.25$ and $j = 2$: planar (**A**), cylindrical (**B**), the planar representation of the cylindrical phyllotaxis (**C**).

The nodes in the cylindrical phyllotactic lattice with $\beta = 0.25$ and $j = 2$ (Figure VIII.6.B) will have coordinates $\left(\pm R, 0, \dfrac{H}{2}M\right)$ and $\left(0, \pm R, \dfrac{H}{4}M\right)$, where R is the radius of the cylinder, H is the step of the ancillary genetic helices, and M is the node number.

If we compare Figure VIII.6.B with Figure III.4.C, we can verify that the lattice at $\beta = 0.25$ and $j = 2$ corresponds to the opposite superposed phyllotaxis.

VIII.7. INTERIM RESULTS

As already described in Section III.1, the most common structures in nature are simple and periodic arrangement structures of primordia, such as distichous, alternate, opposite superposed, opposite decussate, and whorled phyllotactic patterns (Figure III.4). Therefore, most researchers of phyllotaxis have identified these patterns as "primitive" or periodic phyllotaxis, without showing real scientific interest in them.

The analysis of the DH-Model with rational divergence coefficients showed that this "primitive" phyllotaxis can also be described by this model. Moreover, some patterns, in terms of the DH-Model, have a different identification than previously thought. For example, the 3-whorled phyllotactic pattern, which was previously identified as having the divergence angle of 60° ($\beta=0.333(3)$), has the identification $\beta=0.5$ and $j=3$ in the DH-Model. As they say, "nothing is what it seems", especially when analyzing visual phenomena.

In Sections V–VII, the author carried out the DH-Model analysis using the principle from simple to complex — from patterns generated by the Fibonacci series, to patterns generated by generalized recurrent series; from the Archimedean spiral to exponential and logarithmic; from plane to 3D space, so to speak, by induction.

In this Section, the author substituted some rational values with the divergence coefficients ($\beta=0.5$ and $\beta=0.25$) in the DH-Model and obtained patterns previously observed in nature that were identified as "primitive" phyllotaxis. In philosophy, this is called the method of deduction.

It should also be noted that patterns corresponding to phyllotactic lattices with rational divergence coefficients are occurring in nature only in cylindrical phyllotaxis, while lattices with irrational divergence coefficients are occurring in nature in both planar and cylindrical patterns.

DISPUTE OF THE STATICS DH-MODEL

IX.1. NOVELTY AND VALIDITY OF THE DH-MODEL

The universal algorithm for calculating the divergence coefficient for the phyllotactic pattern, the lattice of which is generated by any allowed generalized recurrence sequence, was obtained analytically using mathematical tools of a generalized recurrence sequence, namely, Binet's formula (II.7) and the representation of a recurrence sequence as a sum of recurrence sequences. As a result of the analysis of the minimum distances between the nodes of the phyllotactic lattice, subsets nodes of phyllotactic lattice were identified, called by the author as threads. The analysis of the family of graphs of the dependence of the distance between neighboring nodes on the node number allowed us to identify the ranges within which the distances between neighboring nodes of one thread will be minimal. This made it possible to formalize the concept of "visible parastichy" as a subset of nodes belonging to one thread and whose numbers are within the range, and analytically find the numbers of nodes at the boundaries of these ranges, as well as the numbers of nodes in which the visual effect of "the phyllotaxis rises" occurs. Having drawn the approximating curve through the nodes at the boundaries of the ranges, there it was found that the function of the dependence of the radius of the phyllotactic lattice element (mathematical analogue of primordium) on the node number, which de facto, is a function of the growth of the primordium diameter.

Table IX.1. Comparison of previously known models [3] with the DH-Model.

Previously known models	Novelty in the DH-Model	Validity
Complex spiral phyllotaxis is described by a spiral model with a constant angle of divergence	Both spiral and periodic patterns are described by the DH-Model on a pair of oppositely twisted helices and a constant coefficient of divergence	Known previously as a special case of new results

(Continued)

Previously known models	Novelty in the DH-Model	Validity
The angle between the tangents to the opposed parastichy pair is equal to a right angle in the zone of best visibility of these parastichies	The formula for calculating the angle between tangents to the opposed parastichy pair. According to this formula, the angle between the tangents to the parastichies is equal to a right angle in the node at the boundary of the ranges	Known previously as a special case of new results
The formula for calculating the divergence angle for patterns generated by a recursive sequence 2, 2t+1, 2t+3, 4t+4, 6t+7, ..., where $t \geq 2, t \in \mathbf{N}$	The universal algorithm for calculating the divergence angle for patterns generated by a generalized recurrence sequence $jG_1, jG_2, jG_2+jG_1, 2jG_2+jG_1, 3jG_2+2jG_1, 5jG_2+3jG_1$ with initial terms jG_1 and jG_2, where $j, G_1, G_2 \in \mathbf{N}, 1 \leq j, 1 \leq G_1, 2G_1 \leq G_2$, $\gcd(G_2, G_1) = 1$	Known previously as a special case of new results
The fundamental theorem of phyllotaxis	New proof of the fundamental theorem of phyllotaxis with clarification of the meaning of encyclic numbers	Known previously as a special case of new results
The probability of the appearance of Fibonacci phyllotaxis among spiral patterns is more than 90%	The Fibonacci recurrence sequence, as generating sequence the phyllotactic lattice, does not have "advantages" compared to other allowed recurrent sequence	The DH-Model does not confirm previously known observations

From Table IX.1, it follows that the DH-Model, with a high degree of probability, adequately describes the phenomenon of phyllotaxis observed in nature. The last row of the table is a serious argument against the correspondence of the DH-Model to the natural objects of phyllotaxis. The author believes that the reason for the very high probability of the appearance of Fibonacci phyllotaxis will be found in a dynamic model that analytically describes the process of morphogenesis of the pattern in time.

IX.2. EDGE FUNCTION AND DIVERGENCE COEFFICIENT

As it was defined in Section IV, the planar phyllotactic lattice, in general, is defined by the parametric system (IV.5), which the author called it as the DH-Model. It follows clearly from (IV.5) that the DH-Model contains two mutually independent parameters: the divergence coefficient and the Edge function. The Edge function determines the linear dimensions of the pattern and is visually perceived as the diameters increase of primordia as they move away from the center of the inflorescence. The divergence coefficient determines the angular dimensions of the pattern and is associated with the visual perception of the pattern, through the observation of right- and left-handed spirals (parastichies).

Historically, each of the two independent parameters has several related numerical characteristics of the phyllotactic lattice, which the author divided into two groups: *linear* and *angular*.

- Linear group - numerical characteristics associated with the Edge function: "Plastochrone ratio R", diameter of element of the phyllotactic lattice (primordium);
- Angular group - numerical characteristics associated with the divergence coefficient: divergence angle, the number of right- and left-handed spirals (parastichies), the index of the visible opposed parastichy pair, the angle between the tangents to the opposed parastichy pair.

This statement casts doubt on the well-known stereotype about the relationship of the "Plastochrone ratio R" with the divergence angle [3].

Let's return to the analysis of the parametric system (IV.5) from the point of view of the separation into a linear and angular group of parameters:

$$\begin{cases} x(M) = E(M) \cdot \cos(2\pi\beta M) \\ y(M) = E(M) \cdot \sin(2\pi\beta M) \end{cases}$$

Linear pattern dimensions	Angular pattern dimensions
$E(M)$ is Edge function, which is the distance from the lattice node with number M to the center of the lattice, which determines the shape of genetic spirals. If it is true that the age of primordium with number M is directly proportional to M, then the Edge function depends only on the growth speed of the diameters of primordia.	β is the divergence coefficient, which has a one-to-one correspondence with the generating recurrent sequence. Numerically, this correspondence is expressed in a universal algorithm for calculating the divergence coefficient: $$\begin{cases} |A_1 G_2 - A_2 G_1| = 1 \\ \beta = \dfrac{1}{j} \dfrac{A_2 + \dfrac{A_1}{\tau}}{G_2 + \dfrac{G_1}{\tau}} \end{cases}$$

Linear pattern dimensions	Angular pattern dimensions
	The generating sequence, in the general case, is the allowed generalized recurrent sequence: $jG_1, jG_2, jG_2+jG_1, 2jG_2+jG_1, 3jG_2+2jG_1, 5jG_2+3jG_1$ with initial terms jG_1 and jG_2, where $j, G_1, G_2 \in N$, $1 \le j, 1 \le G_1, 2G_1 \le G_2, \gcd(G_2, G_1) = 1$.

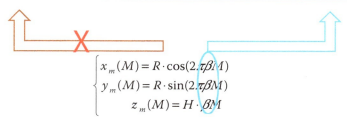

$$\begin{cases} x_m(M) = R \cdot \cos(2\pi\beta iM) \\ y_m(M) = R \cdot \sin(2\pi\beta M) \\ z_m(M) = H \cdot \beta M \end{cases}$$

Unlike planar spiral phyllotaxis, the cylindrical DH-Model is formed by the divergence coefficient only because the cylindrical patterns observed in nature have a constant step of genetic helices. If botanical objects with cylindrical phyllotaxis with a varying helix's step are found, this statement will be revised.

The most important question for understanding the morphogenesis of phyllotaxis is: which parameter in each of the two independent groups is the physical root cause and which is the consequence? The Edge function connects the shape of genetic spirals with the growth speed of primordia diameters. Obviously, the growth speed of primordia diameters is the root cause, and the shape of the genetic spirals is a consequence. The divergence coefficient has a one-to-one relationship with the generating recurrent sequence, but both are a consequence of some physical root cause, finding of which will become possible to analytically describe the process of morphogenesis of the phyllotactic pattern in time. The author believes that this root cause will be found in the dynamic DH-Model.

SOME ASPECTS OF THE DYNAMIC DH-MODEL

The static DH-Model above analysis on a pair (or several pairs) of oppositely twisted helices (IV.1)–(IV.2) showed that almost all known forms of phyllotaxis, ranging from simple periodic to complex spiral patterns, can be described by this model. This allows the author to make an assumption about a single morphological process of formation phyllotactic patterns.

Jean [3] very successfully called the morphogenesis of phyllotaxis as a mathematical puzzle. This puzzle has many facets, each of which needs to be considered separately.

X.1. PHYLLOTAXIS UNDER THE MICROSCOPE OR WHERE IS THE INFLORESCENCE FORMING FROM?

Figure X.1 shows a very early stage of the formation of the pattern of phyllotaxis in the inflorescence of a sunflower. In the center of the inflorescence (a circle with a radius of 2/3 of the radius of the inflorescence), no bulges are noticeable, and in the outer ring with a thickness of 1/3 of the radius of the inflorescence, more than 300 primordia are clearly visible, clearly forming counter parastichies with an index (55,89).

Jean [3] describes the model of phyllotaxis morphogenesis, which is based on the analysis of this micrograph: "Showing the process of floret initiation proceeding toward the center on the generative front with a remarkable degree of symmetry". Ibid: "the primordial florets are initiated in rapid succession from the periphery to the center of the apex"; i.e. under the influence of a certain mysterious mechanism, primordia "initiated" from the outer edge to the center, and the divergence angle is maintained with the highest accuracy and the process itself is not subject to probabilistic deviations.

Despite the obvious fantasticality, this hypothesis has merit that is primordia, which located on the periphery of the inflorescence, is older than those located closer to the center. In the DH-Model terminology, it will sound like this: the age of primordium is directly proportional to the distance from this primordium to the center of the inflorescence. It is then logical to assume that the morphogenesis of the

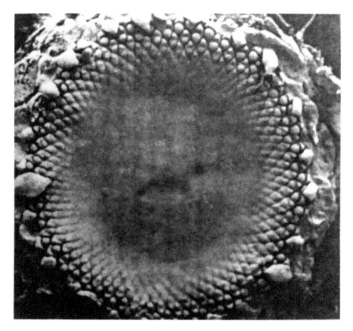

Figure X.1. Micrograph taken by J. H. Palmer with a scanning electron microscope from [3]. Printed with permission from Cambridge University press.

phyllotactic pattern originates from the center of the inflorescence, i.e. each "newborn" primordium appears in the center of the inflorescence and in the process of morphogenesis moves to the periphery. In what follows, let's call the "newborn" primordium as *proprimordium*, and the source from which new proprimordium will appear will be called *a primordial Queen* (similar to a Queen ant).

Following the philosophical thesis "nature loves simplicity", it can be assumed that proprimordium appeared as a single cell, which differs from ordinary cells in that its descendants do not mix with other cells, including descendants of other proprimordia. In the process of mitosis and growth, only one primordium is formed from each proprimordium. A primordial Queen is also a single cell. As a result of the division of a primordial Queen, one of the daughter cells becomes proprimordium, and the other go to a Queen. A similar division mechanism was proposed by Berdyshev [20] (see Appendix **A.2**).

X.2. INVISIBLE PROPRIMORDIUM AND VISIBLE PRIMORDIUM

Just as the birth of a child is a transition from an invisible state to a visible state, a transition from proprimordium to primordium is a transition from an invisible to a visible state. Obviously, this border between the visible and the invisible is very arbitrary and subjective, because it depends on the tool of observation, for example, scanning or transmission electron microscope.

Let's try to explain the process of transition of proprimordium to primordium using a mental experiment. Take a cup with a planar bottom and a set of balls of different diameters. Place the balls in the cup so that each ball will touch the bottom of the cup, then pour an opaque liquid into the cup so that the liquid will cover all the balls. Over time, the liquid will evaporate, and the balls, starting with the largest, will appear above the surface, i.e. become visible. Something similar is happening with primordia: while the proprimordium is very small and located in the receptacle, it is invisible. In the process of growth, the primordium increases and becomes visible, like a tubercle on the receptacle. Video X.1[1] shows a phyllotactic lattice growth model, the invisible part of the primordia is painted in gray, visible in blue. Figure X.2 shows the last frame of this video.

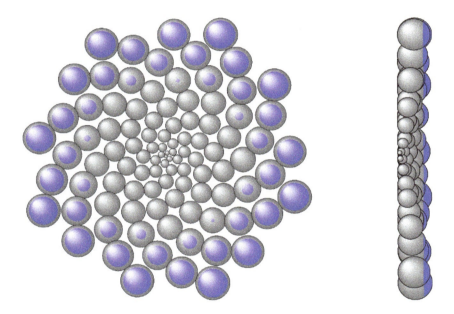

Figure X.2. The last frame of the video of growing model of a phyllotactic lattice.

The hypothesis of an invisible and visible stage of development of primordia well explains the features of the phyllotactic pattern on Figure X.1. The primordia located on the periphery are already quite grown and clearly visible, while those in the center are not yet visible, but already exist. If Palmer [21] could use a transmission electron microscope, then we would see primordia inside the receptacle, as in Figure X.3. As noted above, "nothing is what it seems", especially when analyzing visual phenomena.

---

[1] The video is located at https://youtu.be/5AxPs5Nxxto

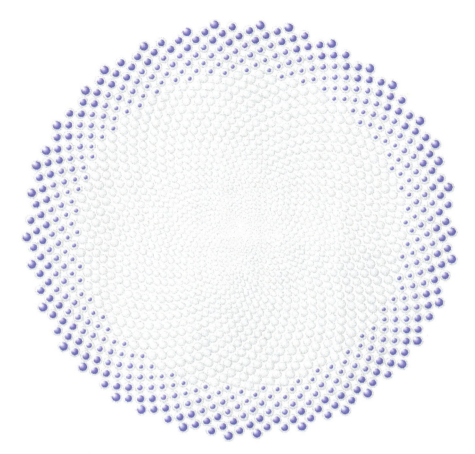

Figure X.3. The phyllotactic lattice for 2000 elements, obtained similarly to the procedure in Video X.1.

The similarity of the phyllotactic lattice in Figure X.3 and micro-photo (Figure X.1) indicates the correctness of both hypotheses:
- the phyllotactic pattern grows from the center of the inflorescence;
- primordium has two stages of development that are invisible and visible.

X.3. THE PHENOMENON OF "CUTTING"

The hypothesis of the invisible and visible stages of development of primordium perfectly explains the phenomenon of cutting out a part of the receptacle [22, 21]. Continuing the analysis of Figure X.1, Jean [3] wrote: "... there is no central area or position with organizing ability," and, "the center of the receptacle appears to have no controlling role since it can be isolated or cut out without changing the floret pattern or rate of floret production."

Dr. Dumais kindly provided Figure X.4 from unpublished [22]. From these microphotographs, it is clearly seen that cutting some part of the receptacle does

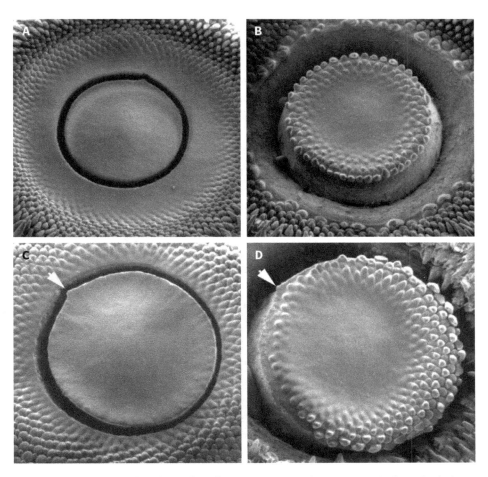

Figure X.4. "Sequential replicas of sunflower meristems after microsurgical manipulation. **A)** and **B)** The same meristem immediately after **(A)** and three days after **(B)** performing a centered cut. **C)** and **D)** another meristem but this time with an off-centered cut. Note in **C)** the inclusion of some patterned tissue in the lower right of the plug leading to unequal development of the plug in **D)**. An arrowhead indicates the same tissue projection in **C)** and **D)**." – from unpublished results [22]. Published with personal permission from Dr. Dumais.

not affect the arrangement of the remaining primordia. This clearly indicates that even before cutting, the proprimordia were already in the receptacle. This confirms the assumption that each primordium is a separate object, developing according to an autonomous plan, so the removal of a certain number of proprimordia does not affect the growth and location of the others.

The second important conclusion from the analysis of these microphotographs: the formation of the pattern of phyllotaxis takes place during the invisible stage of development of primordia. Therefore, by the time the early primordia become visible, the structure of the phyllotactic pattern has been already formed.

X.4. THE TUBE MODEL OF THE PERIODIC PHYLLOTAXIS

Periodic (non-spiral) forms of phyllotaxis, as discussed in Section VIII, are not only the most common in nature, but also the simplest from a mathematical point of view.

The analysis of periodic patterns shows that in nature, they are found exclusively in cylindrical form. This suggests that the morphogenesis of periodic phyllotaxis occurs in some internal space, which is bounded by a cylinder; we will say in the tube.

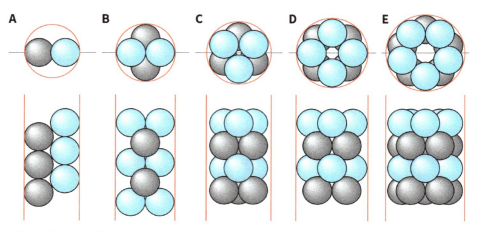

Figure X.5. **A** is distichous or alternate phyllotaxis ($\beta = 0.5, j = 1$), **B** is opposite superposed phyllotaxis ($\beta = 0.25, j = 2$), **C** is 3-whorled phyllotaxis ($\beta = 0.5, j = 3$), **D** is 4-whorled phyllotaxis ($\beta = 0.5, j = 4$), **E** is 5-whorled phyllotaxis ($\beta = 0.5, j = 5$).

Let's suppose that the primordia were "born" from the primordial Queen. Immediately after their appearance, they are pushed into a tube, diameter of which is slightly larger than the diameter of the primordia. The packaging of proprimordia in the tube occurs with maximum density, so the mutual disposition of proprimordia will depend on the ratio of the diameter of the primordium d_{pr} and the diameter of the tube D_{sr} (Figure X.5).

		β	j	min d_{pr}/D_{sr}	max d_{pr}/D_{sr}
A	distichous or alternate	0.5	1	1	2
B	opposite superposed	0.25	2	2	2.15
C	3-whorled	0.5	3	2.15	2.41
D	4-whorled	0.5	4	2.41	2.7
E	5-whorled	0.5	5	2.7	3

If the velocity of appearing proprimordia is much higher than the velocity of the growth of their diameter, then during the growth process, there are no visible distortions of the periodic structure of phyllotaxis and primordia visually seem the same (appeared simultaneously).

X.5. THE MATHEMATICAL GENESIS OF THE FIBONACCI NUMBERS OR WHY TURING HAD FAILED IN HIS RESEARCH IN PHYLLOTAXIS

In Sections IV–VIII of this research, the author tried to answer the philosophical question: "What is a visible spiral in a sunflower?" that was set out in the Introduction. Now, let's return to the adult question: "Why is the number of right- and left-handed spirals equal to a pair of neighboring numbers from the recurrent sequence?"

As it is known, a plant is a biological object, consisting of a large number of cells, between which various physicochemical processes occur. These processes are adequately described by statistical and differential-integral mathematical tools. Therefore, most researchers tried to build a model of phyllotaxis morphogenesis, relying on mathematical methods that were well tested in biology and physics.

The most prominent representative of this direction was the outstanding mathematician of our time — Alan Turing, who was one of the fathers of cybernetics and who built a computer that cracked the code of the Nazi cryptographic machine Enigma, which made a huge contribution to the victory of the Allies in WW2.

In the early 1950s, Turing published the fundamental article "The Chemical Basis of Morphogenesis" [14], devoted to the self-organization of matter and the self-oscillating chemical reactions. At the same time, Turing was interested in the phenomenon of phyllotaxis and tried to explain it using the same physicochemical methods. In 1992, Turing's unfinished article "Morphogen theory of phyllotaxis" [6, 7] was published, in which he tried to explain the appearance of Fibonacci parastichies by the self-oscillation nature of chemical reactions in plants. However, the Fibonacci numbers do not arise in solving equations describing self-oscillating processes. The Fibonacci numbers also do not appear as basic constants in statistics or in the differential-integral calculus, such as the constant π arises in trigonometry or e in the differential-integral calculus.

The author hypothesizes that the mathematical tool that describes the phenomenon of phyllotaxis should contain the Fibonacci numbers as a basic constant. Therefore, it is necessary to consider the mathematical genesis of the Fibonacci numbers, i.e. find a circle of mathematical problems, the solution of which gives rise to the Fibonacci numbers as a basic constant.

It is believed that the first mention of the Fibonacci numbers is "The Rabbit problem" from the treatise "Liber abaci" (1202) of the outstanding mathematician of the Middle Ages, Leonardo Pisano, is better known by his nickname Fibonacci, which means "son of Bonacci".

Leonardo Pisano thus formulated the task (more precisely, he translated Indian mathematicians from the Arabic language): let there be a newborn pair of immortal rabbits (male and female) in a separate place. Two months later, this pair beget a new pair of immortal rabbits and continues to beget a new pair every month. Each new pair of rabbits, also after two months, beget a new pair of rabbits and continues to beget new pairs each month. Question: how many pairs of rabbits will there be in n months?

From the solution of the task, it follows that the number of pairs of rabbits in the n-th month will be equal to the sum of the pairs of rabbits in the $(n-1)$-th and $(n-2)$-th months. That is, the already known recurrence formula (II.3):

$$F_n = F_{n-1} + F_{n-2}$$

It should be noted that any recurrence sequence with arbitrary initial terms can be expressed in terms of the Fibonacci sequence:

$$G_n = G_{n-1} + G_{n-2} = F_{n-2}G_2 + F_{n-3}G_1$$

Therefore, the Fibonacci sequence is the basic constant for all recurrence sequence.

Another "source" of the Fibonacci numbers is Pascal's triangle - an infinite table of binomial coefficients $\binom{n}{i}$, which has a triangular shape. Pascal's triangle has a one at the top and two ones in the second row (Figure X.6). Starting from the third row, each number is equal to the sum of two numbers located above it.

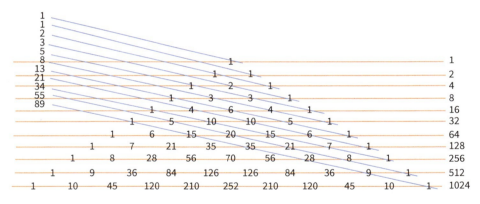

Figure X.6. Pascal's triangle.

Let us recall that binomial coefficients are generated from the expansion of Newton's binomial:

$$(a+b)^n = \sum_{i=0}^{n} \binom{n}{i} a^{n-i} b^i,$$

where $\binom{n}{i} = \dfrac{n!}{i!(n-i)!}$ is the binomial coefficient, and $n! = 1 \cdot 2 \cdot 3 \cdot \ldots \cdot (n-1)\, n$ is the factorial of a number n.

If we add the binomial coefficients in the n-th horizontal row (brown lines in Figure X.6), then their sum will be 2^n. Indeed, if $a = b = 1$, then

$$2^n = (1+1)^n = \sum_{i=0}^{n}\binom{n}{i}$$

If we calculate the sum of the binomial coefficients along the diagonal (blue lines in Figure X.6), then the sum will be equal to one of the Fibonacci numbers:

$$\sum_{i=0}^{n/2}\binom{n-i}{i} = F_n$$

It should be noted that the binomial itself can be represented as a recursive procedure:

$$(a+b)^n = (a+b)^{n-1} \cdot (a+b)$$

It follows that the Fibonacci numbers are generated in Pascal's triangle by recursive procedure too. It is interesting that the exponential series 1, 2, 4, 16, ..., 2^{n-1}, 2^n can also be determined using the recursive formula:

$$D_n = D_{n-1} + D_{n-1} \text{ with } D_0 = 1$$

$$2^n = 2^{n-1} + 2^{n-1}$$

As we know, numbers in computers are represented in binary code, i.e. the coding is based on the exponential series 2^n. Based on the fact that both the 2^n series and the Fibonacci series are both recursive and connected through Pascal's triangle, Stakhov [23, 24] substantiated the possibility of using the Fibonacci series to coding numbers in computers (more details are described in Appendices **B** and **C**).

Mathematical problems are also known whose optimal solution is the Fibonacci numbers:

- The optimal set of weights for the solution of the Bachet-Mendeleev problem (weighing with delay), Stakhov [25, 23];
- Optimal algorithms for searching and sorting data, Knuth [26].

It is interesting that the solution of the Bachet-Mendeleev problem about weighting without delay (ideal case) is a set of weights: 1, 2, 4, 16, ..., 2^{n-1}, 2^n. Also, one of standard algorithms for searching and sorting data use binary search.

It should be noted that both the weighing and the data sorting are recursive processes; that is, at each step of the process, some operation is performed on the results obtained in the previous step.

As can be seen from the examples above, the 2^n sequence is the basic constant of some ideal recursive processes. For example, if in the rabbit problem each pair would give birth to a new pair, starting from the first month, then the number of pairs of rabbits would be 2^n in the n-th month.

In contrast, the Fibonacci sequence is the basic constant of some *not ideal* recursive processes that have an internal delay or shift. This is intuitively evident from a comparison of recurrence formulas:

$$D_n = D_{n-1} + D_{n-1}$$
$$F_n = F_{n-1} + F_{n-2}$$

From what was said in this chapter, one can make the assumption that recursive processes have at least two basic constants: the 2^n series and the Fibonacci sequence.

X.6. THE CONCEPT OF THE DYNAMIC DH-MODEL

The dynamic model of the phyllotactic lattice growth assumes a transition from a phyllotactic lattice with N elements to a lattice with $N+1$ elements, herewith:

- in the phyllotactic lattice, the structure of genetic spirals (helices) and a constant divergence angle are saved;
- each element grows (increases in size);
- each element moves away from the center;
- at regular intervals, a new element is added to the center of the lattice.

The time between the appearance of new elements will be called *the growth tact*.

It should be noted that the growth of each element and its movement from the center occur continuously and uniformly, and the addition of a new element is discrete.

From the abovementioned, it follows that this dynamic model is a recursive process. In mathematics, recursive statements are proved by the method of mathematical induction, which consists of two points:

1. Proofs of the truth of the statement for the first step;
2. Proofs of the truth of the statement for the transition from the N-th to the $(N+1)$-th step.

X.7. THE HYDRAULIC ASPECT OF THE DYNAMIC DH-MODEL

In order to remain the phyllotactic lattice structure constant during the transition from the phyllotactic lattice with N elements to the lattice with $(N+1)$ elements, it is necessary for every i-th element in one growth tact:

- becomes equal in size to the $(i+1)$-th element at the beginning of this tact;
- moves from the center so that the distance from the center to the i-th element becomes equal to the distance from the center to the $(i+1)$-th element at the beginning of this tact.

A logical question arises: how does the new element, the smallest of all the elements of the lattice, "push apart" the entire lattice so that the lattice does not lose its structure?

As suggested in Section V.6, the phyllotactic lattice element is like an incompressible and deformable silicone disk, whose center coincides with the lattice site and such elements of the lattice fill out all the space of the phyllotactic lattice. The combination of these elements, in their mechanical properties, will be similar to a single piece of silicone. Therefore, to describe the interaction of elements from such a combination, hydraulics is the most suitable.

Each i-th element, except those located on the periphery of the phyllotactic lattice, is surrounded by other elements. In terms of hydraulics, the growth of an element in a fluid-like environment, creates pressure on this environment. The value of this pressure will be proportional to the increase in the area of the element.

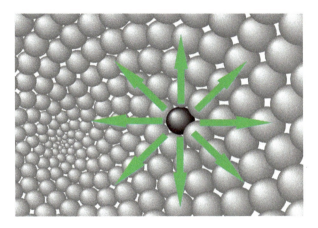

Figure X.7. A pressure distribution in the hydraulic model.

As in the analysis of the static model, we restrict ourselves to consider the planar Archimedean Fibonacci phyllotaxis. As it has been proven in Section V, the diameter of the lattice element of the planar Archimedean phyllotaxis is proportional to the square root of the number of this element and does not depend on the generating recursive series (formulas V.6.4, V.9.13, V.10.20, and V.12.14):

$$d_i = \sqrt{2\pi i}$$

where i is the element number. We also use the circle area formula:

$$S_i = \frac{\pi}{4} d_i^2$$

to calculate the pressure created by the i-th element in one growth cycle,

$$\Delta\rho = k(S_{i+1} - S_i) = \pi k(r_{i+1}^2 - r_i^2) = 2\pi^2 k(i+1-i) = 2\pi^2 k, \tag{X.1}$$

where **k** is a constant coefficient. From formula (X.1), it follows that each lattice element of the planar Archimedean phyllotaxis creates the same pressure, regardless of its current size.

According to the laws of hydraulics, pressure is distributed evenly in all directions, i.e. each element produces pressure on all other elements (Figure X.7). On the other hand, all other elements press on this element.

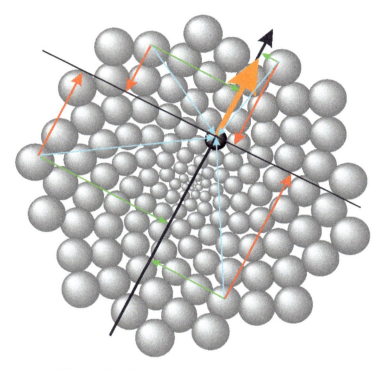

Figure X.8. The pressure distribution in the hydraulic model of the phyllotactic lattice.

Consider the interaction of an arbitrary i-th element (black circle in Figure X.8) with the totality of the remaining elements of the phyllotactic lattice. Let's draw two mutually perpendicular lines. The first line (the thick black line in Figure X.8) passes through the center of the phyllotactic lattice and the center of the i-th element – let's call it *the center line*. The second (thin black line) passes through the center of the phyllotactic lattice and is perpendicular to the center line.

As you know, pressure (or force) has a direction, which is a vector quantity, and a vector can be represented as a sum of two vectors.

The pressure vector of an arbitrary k-th element on the i-th element (cyan color arrow) can be represented as the sum of two perpendicular vectors — red and green.

The red vector is parallel to the central line, and the green is perpendicular to this line. The sum of the green vectors, from all the elements, will be zero because the elements of the phyllotactic lattice are located almost symmetrically relative to the central line. Whereas the sum of the red vectors from all the elements will be directed strictly from the center of the lattice to the center of the i-th element (orange arrow in Figure X.8). Under the influence of this pressure, each element of the phyllotactic lattice will move from the center, and the entire lattice will move apart without violating its structure. Such an expansion of all elements of the lattice frees up space for a new element.

Video X.2[2] shows the phyllotactic lattice growth process with the growth tact of one second. Every second, with the advent of a new primordium in the center of the lattice, the phyllotactic lattice transitions from the N-th to the $(N+1)$-th condition.

X.8. EXPERIMENTAL VERIFICATION OF THE HYDRAULIC MODEL

An experiment with physical objects along with computer simulation is necessary to confirm or refute any hypothesis. This experiment must be repeated by another researcher, in a different place and with a different set of equipment.

The experimental verification of the hydraulic model proposed above is a very difficult task because, in order to conduct such an experiment, it is necessary to find a physical object that will have properties similar to a growing primordium, namely to the constant rate of diameter increase and the ability to start an increase in diameter at a certain point of time.

Finding or creating such a physical object is very difficult. Therefore, the author suggests to find a non-linear curvature of 3D space when applying to the planar spiral phyllotactic lattice, a 3D lattice will be obtained, all elements of which will have the same size, and the lattice will retain the structure of visible parastichies. For brevity, we will call such the curvature of the 3D space as *R-transformation*. The result of R-transformation will be a 3D lattice consisting of the similar elements that human perception can combine into spirals, which are commonly called parastichies. Also, we will call such a 3D structure as *R-lattice*. R-transformation is similar to transforming the planar spiral phyllotactic lattice into a lattice on a cone from Video VII.1.

To prepare for the R-transformation, it is necessary to mentally replace each cylindrical element of the planar phyllotactic lattice, from Section V.6, with a ball of the same diameter. Obviously, the centers of all balls will lie on the same plane. R-transformation will be subject to a limited number of lattice elements, from zero to N.

Two conditions must be satisfied in the R-lattice:

– all elements (balls) of the R-lattice must have the same diameter equal to the diameter of the largest ball before the R-transformation, $D = \sqrt{\pi N}$ according to (V.6.4);

---

[2] The video is located at https://youtu.be/GRrN-aoH09k

– balls, which are in contact in the planar model before the R-transformation, will touch the external surfaces after the R-transformation.

The process of R-transformation of a planar phyllotactic lattice into an R-lattice can be seen on Video X.3[3].

On the R-lattice, which is shown in Figure X.9, one can see parastichies with indices 2, 3, 5, 8, 13, and 21 without any special eye strain. Human perception cannot distinguish such a large number of parastichies with different indices on either the planar or cylindrical phyllotactic lattice.

The author made the necessary mathematical calculations and calculated the shape of the funnel, which passes through the centers of the elements of the R-lattice. The axial section of this funnel is indicated in red in Figure X.9 and Figure X.10. On the 3D-Object[4], you can see the 3D-model of the R-lattice, the centers of the elements (balls) of which lie on the funnel.

Using identical and invariable physical objects, such as plastic or silicone balls, to simulate the elements of the R-lattice, makes it possible to conduct a physical experiment that will confirm (or refute) the hydraulic model of the morphogenesis of phyllotactic patterns.

To implement the experiment, it is necessary to manufacture a device whose axial section is shown in black in Figure X.10, which we will call *Y-funnel*. The Y-funnel consists of an external and internal funnel between which there will be balls simulating elements of the R-lattice.

The diameter of the top edge of the Y-funnel will be $N + \sqrt{2\pi N}$, and the diameter of the bottom will be equal to the diameter of the ball. Modern 3D printing technology allows you to make Y-funnel from a transparent material with a high degree of accuracy.

For the experiment, an empty Y-funnel is mounted vertically, as in Figure X.10. Then, the balls slowly and evenly move into the Y-funnel., The balls, that are already in the Y-funnel, will move upward under pressure from the balls that enter the Y-funnel from below. These balls will fill the space between the outer and inner surfaces of the Y-funnel. If the hydraulic model is correct, then the balls will form visible parastichies as it is shown in Figure X.9.

As it was noted at the beginning of this section, to confirm the objectivity of the experiment, it should be repeated by other researchers who will have a different set of equipment. Therefore, to maintain the purity of the experiment, interested readers are invited, on one's own, to find mathematical functions that will describe the Y-funnel, and also, build a Y-funnel and conduct their own experiment.

---

[3] The video is located at https://youtu.be/GSxyVoFgpWk

[4] The 3D Object is located at https://1drv.ms/u/s!An67HioxMGoVxCHrKiNqw_BvG-1oh?e=1lb0Np

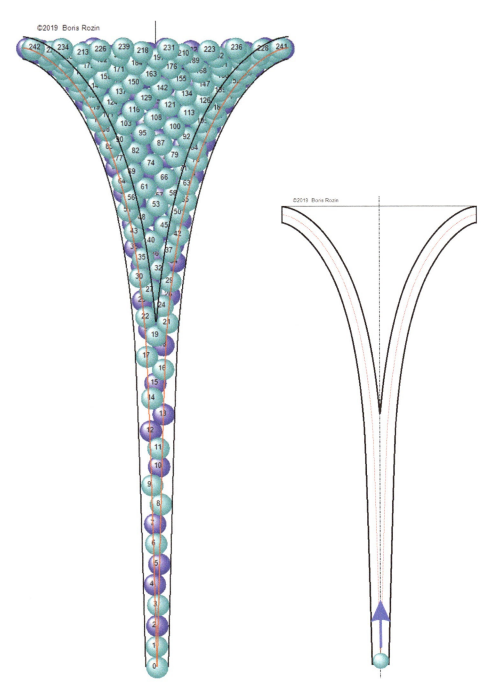

Figure X.9. The R-lattice.

Figure X.10. The axial section of the Y-funnel.

WHAT WE OBJECTIVELY KNOW AND DON'T KNOW ABOUT PHYLLOTAXIS

The transition from a static model to a dynamic one requires an audit of our knowledge of phyllotaxis.

As noted in the Introduction, the history of mathematical phyllotaxis is 250 years old. Over the years, hundreds of researchers, mostly mathematicians and botanists, have tried to uncover the mystery of "Fibonacci spirals." However, the variety of geometric shapes, difficulties of the measurement, and subjective perception in the identification of phyllotactic patterns have generated many mutually exclusive models and false stereotypes. Even the meaning of the term *phyllotaxis* is widely interpreted by various researchers. Let remind that the author defined phyllotaxis as a visual perception of the spatial structure of primordia (Section IV).

In order to get out of the labyrinth of subjective perception, it is necessary to follow the thesis of Immanuel Kant [27]: "a doctrine of nature can only contain so much science proper as there is in it of applied mathematics."

Even 30 years ago, the main type of the measurement of phyllotaxis was the identification of a generating recurrent sequence by counting the index of opposed parastichy pair. That is, the main attention was paid to the "angular" group of parameters. Accordingly, the main analysis method was to calculate the frequency repetition (spectrum) of the generating recurrent sequences. The lack of accurate tools for direct measuring some other parameters of the phyllotactic pattern, (for example, the shape of the genetic spirals or the "Plastochrone ratio R"), gave a rise to false stereotypes.

Nowadays, the most complete collection of numerical data on phyllotaxis can be seen in [3]. Jean did a great job of sorting and compiling the observation data of spiral phyllotactic patterns by many researchers who used different techniques.

According to the author, despite the large volume, the data on the distribution of repetition frequencies (spectrum) of the generating recurrence series from [3] have significant drawbacks:

- it is impossible to verify (repeat observations) these phyllotactic patterns due to the fact that these observed plant instances are no longer available for analysis;
- the probability of the appearance of non-Fibonacci patterns is on the verge of statistical error.

In confirmation of the last point, an interested reader can independently try to find a plant with non-Fibonacci spiral phyllotaxis in a botanical park. It is also difficult to find a photo of a non-Fibonacci pattern on the Internet.

According to [3], the relative amount (4%–6%) of non-Fibonacci spiral patterns are approximately equal to the percentage of the defects in semiconductor manufacturing. If we assume that non-Fibonacci spiral patterns are defective patterns of Fibonacci patterns, then the proof or refutation of this assumption will have a decisive influence on our understanding of the morphogenesis of phyllotaxis. Therefore, non-Fibonacci spiral patterns require a particularly detailed measurement and analysis.

It should also be recognized that the Bravais-Bravais theorem and the fundamental theorem of phyllotaxis, which are the basis of mathematical phyllotaxis, exist in the framework of a spiral model with a constant divergence angle, and the model itself is only the result of a scientific assumption.

Following the thesis of Immanuel Kant presented above, only a qualitatively new level of phyllotaxis measurement will allow to obtain qualitatively new results in the study of the morphogenesis of phyllotaxis. Therefore, to measure phyllotactic patterns, modern digital technologies (Artificial Intelligence and pattern recognition) are necessary.

HOW PHYLLOTAXIS SHOULD BE MEASURED

A qualitatively new level of phyllotaxis measurement can be achieved using digital technologies, Artificial Intelligence, and a network organization for collecting raw data.

XII.1. DIGITAL TECHNOLOGY AND ARTIFICIAL INTELLIGENCE IN THE MEASUREMENT OF PHYLLOTAXIS

The digital measurement algorithm of phyllotaxis is based on the following statements:

- The index of visible parastichies is a secondary parameter, and the primary is the relative arrangement of primordia (lattice nodes) in the inflorescence;
- As follows from the DH-Model, in the formation of the phyllotactic lattice, two independent groups of parameters participate - "linear" and "angular", which must be measured separately;
- The use of digital technologies, Artificial Intelligence, and pattern recognition algorithms allows a more accurate measurement of the relative position of primordia on the phyllotactic pattern.

This allows us to formulate a basic algorithm for the digital measurement of phyllotaxis:

1. Digitization of raw data:
   a. Loading a graphic (video or 3D) file with the phyllotactic pattern;
   b. Digitization of the center of primordia;
   c. Saving the coordinates of the center of primordia (nodes of the phyllotactic lattice).
2. Sorting nodes by distance from the center:
   a. Calculation of the coordinates of the center point (or axis for cylindrical phyllotaxis) of the phyllotactic lattice;

b. Calculation of the distance from each node to the center of the lattice;
   c. Sorting nodes by increasing the distance to the center of the lattice;
   d. Numbering nodes based on sorting results.
3. Finding the best correlating functions for parameters from "linear" group:
   a. $E(M)$ — Edge function (form of genetic spirals);
   b. "Plastochrone ratio R";
   c. Dependence of the radius of the phyllotactic lattice element on the node number.
4. Calculation of parameters from the "corner" group:
   a. The average divergence angle (coefficient);
   b. Dependence of fluctuations of the divergence angle on the node number;
   c. The index of opposed parastichy pair.

At the pilot stage of this project, the raw data for measuring will be in high-resolution digital photos with phyllotactic patterns. The center of primordia will be digitized manually from a high-resolution digital photo in a special software application (1.b). In the next step, this process will be automated with recognition algorithms. In the third stage, for measuring convex and cylindrical phyllotactic objects, 3D scanning or software will be used in order to convert a video into a 3D model.

The author wrote the code for a pilot program in C# language that implements the algorithm. During this software testing, it was found that the measurement errors (inaccuracy in determining the center of primordia) and natural deformations on the real botanical objects, when doing it critically, affect the accuracy of calculating the center of the inflorescence and invalidate the results of sorting nodes. Due to this reason, the algorithm should be supplemented with an Artificial Intelligence unit for analysis and adjustment of the coordinates of the centers of primordia. In the Artificial Intelligence unit, the phyllotactic lattice will be described as a connected graph in which the nodes of the lattice are the vertices of the graph and the distances between the nodes are the edges of the graph. Knots belonging to different threads (parastichies) can be found by a method similar to the well-known the traveling salesman problem, which will allow you to adjust the coordinates of the centers of primordia and renumber the nodes.

XII.2. THE NETWORK PROJECT "OPEN PHYLLOTAXIS"

In order to obtain reliable data, it is necessary to measure and analyze many thousands of biological objects. The author offers a network project called Open Phyllotaxis. The project "Open Phyllotaxis" will become a virtual platform for collecting, storing, analyzing, demonstrating, and discussing all aspects of the phenomenon of phyllotaxis. The project's ideology consists in the openness of the primary data (phyllotactic patterns photos), free access to the measurement and analysis of results, as well as the possibility of using other algorithms and programs to measure and analyze primary data by those interested in the subject researchers.

The second feature of the project will attract high school and college students to the Clubs of Open Phyllotaxis network. The main task of these clubs will introduce students to the phenomenon of phyllotaxis and collect primary data. Moreover, participating in the project does not require knowledge beyond the framework of the high school curriculum.

Each member of the local Club of Open Phyllotaxis will be a member of the entire project and will be able to share the photos of phyllotactic patterns to the server, where the photos will be available for analysis.

Particular attention should be paid when collecting the photos containing non-Fibonacci spiral phyllotactic patterns. The result of the project will be the accumulation of primary photos and the measurement results of phyllotactic patterns. This will allow us to make many new discoveries in phyllotaxis, as well as to check the degree of correlation of the DH-Model and other models with the real biological objects. At the same time, due to the collection of the photos for analysis, an online museum-gallery of phyllotactic patterns could be created.

In addition to the scientific, this project has also an enormous educational potential. As the author has written in the Introduction, phyllotaxis is one of the few phenomena of nature in which biology and mathematics are clearly combined. Young participants of the project "Open Phyllotaxis" will be able to see the beauty in mathematics and the mathematics in beauty. Even indirect participation in Open Phyllotaxis network will give high school and college students some experience in the real scientific research. As the experience of such programs will show, this is a great incentive for learning and doing science in the future.

XIII

CONCLUSION OR FIVE CRITICAL QUESTIONS

In this conclusion, the author briefly recalls the methodology and the main stages of this study.

The search for the answer to the children's question "why do we see spirals on inflorescence of a sunflower?" led to the hypothesis about the visual matter of spirals on the phyllotactic pattern. Understanding of that parastichies are a visual effect which led to the idea of calculating the distances between nearby nodes of the phyllotactic lattice. The analysis of these distances became the basis of the DH-Model research.

At the first stage, the static DH-Model was considered in detail for the most typical form of phyllotaxis that is the planar Archimedean Fibonacci lattice. The analysis of the distances between nearby nodes of the lattice allowed the author to determine the key concepts of the DH-Model: a thread, a family of threads, a diapason, a border of diapason, and the node of "the phyllotaxis rises". The family of graphs of the distances between neighboring nodes for threads of all families, based on this analysis, made it possible to find the dependence of the diameter of the lattice element on its number and to understand mathematical basis of the visual perception of the phyllotactic pattern. Also, there was analytically found the function of the dependence of the angle, between the tangents to the opposite parastichies, on the number of the node at which these parastichies intersect. This confirmed the previously found intuitive rule: in the zone of the best visibility, the tangents to the pair of visible parastichies intersect at a right angle.

Furthermore, the same analysis sequence was applied to planar Fibonacci lattices generated by Fibonacci-like, non-multiple, and generalized recurrence sequence. There, it was proven by the author that the converse of the fundamental theorem of phyllotaxis and analytically found the universal algorithm for calculating the divergence coefficient, which is valid for all allowed generating recurrence sequence. This allowed us to conclude that the DH-Model is adequate for all planar phyllotactic lattices.

The comparison of DH-Models with various forms of the planar genetic spirals showed that only power spirals (the Archimedes spiral is a particular case), and not logarithmic ones, can be genetic spirals for patterns with an "rise" effect.

An analysis of the distances between nearby nodes applied to cylindrical lattices generated by Fibonacci, multiple, and generalized recurrence sequences proved the validity of the universal algorithm for calculating the divergence coefficient for all allowed generating recurrence series. This allowed us to conclude that the DH-Model is adequate for cylindrical phyllotactic lattices.

The analysis of the DH-Model for cylindrical phyllotactic lattices with various rational divergence coefficients showed that periodic patterns can also be described by this model.

A review of the internal structure of the DH-Model showed that all the parameters of this model can be divided into two groups: "linear" and "angular". The "linear" group includes the parameters associated with the Edge function: "Plastochrone ratio R" and the diameter of the element of the phyllotactic lattice. Whereas the "corner" group includes parameters related to the divergence coefficient: divergence angle, the number of right- and left-twisted spirals (parastichies), the index of the visible opposed parastichy pair, and the angle between the tangents to the opposed parastichy pair.

A comparative analysis of the DH-Model as a recursive structure and experiments on cutting out primordia at a very early stage made it possible to draw an unambiguous conclusion that the inflorescence grows from its center and the formation of the phyllotaxis pattern occurs much earlier than the primordia becomes visually distinguishable. This made it possible to formulate and prove the theoretical possibility of the hydraulic DH-Model of the inflorescence morphogenesis.

The comparison of the obtained results with the results of other researchers, primarily with the data generalized in [3], in addition to numerous confirmations, revealed a few discrepancies. However, to date, there is an insufficiency of detailed measurements of the real botanical objects for an unambiguous conclusion in favor of these or other models. To achieve a qualitatively new level of phyllotaxis measurement, the author proposed an algorithm for digital phyllotaxis measurement using Artificial Intelligence and a primary data collection system, which he called the network project "Open Phyllotaxis".

An experimental verification of the hydraulic model and the implementation of the network project "Open Phyllotaxis" will provide data for analysis that will answer **Five critical questions** that do not have unambiguous clear answers today:

1. Are non-Fibonacci spiral patterns stably repeating structures or are they defective instances of Fibonacci patterns?

2. What is the shape of the genetic spirals of the real botanical objects (the Archimedean, logarithmic, exponential or another spiral)?
3. Is there a relationship between the "Plastochrone Ratio R" and the divergence angle?
4. Is the DH-Model adequate to the real botanical objects?
5. Is the hypothesis of the hydraulic model of the phyllotactic pattern morphogenesis true?

And the last question of this book: why do we study phyllotaxis?

The knowledge of the mechanism of morphogenesis of phyllotactic patterns will allow us to come closer in understanding the morphogenesis of the living nature, including the human body. This knowledge will allow people to find qualitatively new ways of health treatment and protection of human beings for many years ahead.

APPENDIX

THE GOLDEN SECTION AND MORPHOGENESIS

The ancient Greeks, who had a great influence on modern culture, considered the Golden Ratio to be the canon of beauty and world harmony. They believed that harmony was present both in the person himself and in the fact that the person manufactured.

Every year, dozens of books devoted to the Golden Section, are published in various publishing homes, and the number of websites that mention this phenomenon is incalculable. The authors of these "studies" find more and more manifestations of the Golden Section in various aspects of the world around us. However, a closer look reveals only a few areas in which the presence of the Golden Section is confirmed by an objective measurement:

- Phyllotaxis (angle of divergence) [1, 3, 4]
- Proportions of the human body [28, 29]
- Architecture and sculpture [30, 31]
- Painting, housewares, especially antique [32, 33]

Many researchers have tried to unravel the mystery of the Golden Section. The author did not escape this temptation either. While he was an university student, the author proposed a hypothesis, which was published in Russian in 1991 [34] and 1993 [35], and then in 2003 [36] in English.

As the author can sees it, visual environment can be divided into two points of view: objects that man created and things which were formed by nature.

A.1. THE GOLDEN RATIO IN THE WORKS OF MAN

The presence of the Golden Proportion in the forms of objects created by human beings is possible to explain by means of the analysis of the famous researches:

- In Fechner's experiments [37], the examinees were asked to choose the most "beautiful" rectangle from a set, from a square to a double square. The overwhelming majority selected the rectangle in which the correlation of the sides

was τ. Therefore, subjects whose forms contain the Golden Proportion are perceived "favorably." Not by accident, the credit cards have the correlation of length and width equal to the Golden Proportion.
- The research works by Zeising [28], Hambidge [38], and Dochi [29] indicated the presence of the Golden Proportion in a correlation of the body parts of man (Figure A.1), specifically the hand.

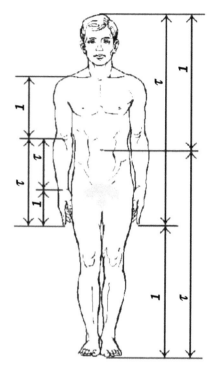

Figure A.1. The Golden Proportion in a human body.

From the studies of Fechner and Zeising, it follows that the Golden Section is present in a human body at least twice: in a favorable perception and the proportions of the body. Therefore, if a person wants to create something beautiful, then he is intuitively guided by his favorable perception and the standard of the Golden Proportion — the human hand, which he always has "ready at hand." Evidently, the outcome of the creation will be an object that contains the Golden Proportion.

A.2. THE GOLDEN SECTION IN NATURE

To explain the presence of the Golden Section in the forms of the living nature, the author suggested that the Fibonacci numbers and the Golden Section were mathematical descriptions of a certain morphologic process. On a micro level (integer-valued level), the quantitative characteristic of this process is shown as the

The Golden Section and Morphogenesis **137**

Fibonacci numbers, and on a macro level (statistical level), it is shown as the base of the Golden Proportion — the number τ.

If such a morphological process is the law of living matter, then it is possible to explain the presence of the Golden Proportion in the correlation of the parts of the human and animal bodies by means of this law, and it's also possible to explain the phenomenon of phyllotaxis.

Nowadays, a scientific opinion prevails that each cell of the body is divided into two identical cells (mitosis or symmetric division), and both newborn cells continue to divide in the same way. In [36], Rozin suggested that the two cells, which were formed as a result of the division that were not the same. One of these cells missed the first cycle of division. Rozin [36] called such a morphological process F-division or asymmetric division.

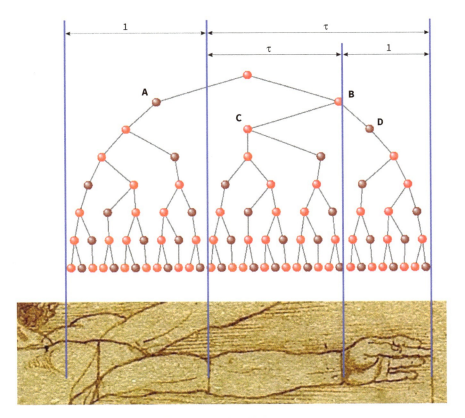

Figure A.2. The F-division.

Let us consider the quantitative characteristics of the F-division. Thus, after the first cycle of the F-division, the two cells **A** and **B** will be produced (Figure A.2), although only **B** will be divided during the second cycle. After the two cycles of the F-division, three cells will result, where only two cells of which will be divided during the third cycle. After the third cycle, the total amount of cells will equal to five; only three cells of which will be divided during the fourth cycle of the F-division, etc. Therefore, during the F-division, there will be 2, 3, 5, 8, 13, 21,... cells produced from the first one.

This is the famous the Fibonacci sequence, which is calculated by the recurrence formula (II.3).

Table A.1. The number of cells in the process of F-division.

Number of cycle division		0	1	2	3	4	5	6	7	n
Total cells	from the initial cell	1	2	3	5	8	13	21	34	F_n
	from cell **A**		1	1	2	3	5	8	13	F_{n-2}
	from cell **B**	1	1	2	3	5	8	13	21	F_{n-1}

As it can be seen from Table A.1, from cell **B,** there will be more cells in τ than from **A**.

This hypothesis about the F-division of cells may explain the presence of the Golden Proportion linear dimensions of the human body, for example, in the arm (Figure A.2).

For example, let's at a certain stage of development of the embryo, select one cell from which a hand will develop. After the first F-division, two cells **A** and **B** will be formed (Figure A.2). Cell **A** will miss the next division cycle; therefore, the numbers of its descendants will be in τ times smaller than the cells generated by **B**. Indeed, as can be seen in Figure A.2, the ratio of the sum of the lengths of the hand to the length of the elbow is the Golden Section. If the length is proportional to the number of cells, then from cell **A,** the forearm will develop, and from cell **B,** the hand and the elbow will develop. Similarly, after dividing cell **B**, cells **C** and **D** are formed. From the descendant cell **C**, an elbow is formed, and from the descendant cell **D,** a wrist, etc. to the phalanges of the fingers on the hand.

In mitosis (symmetric division), the aggregate of embryonic cells forms a symmetrical and equilibrium structure in 3D space. In contrast, during F-division, the structure of the embryo will be asymmetric and not equilibrium.

Rozin [36] believes that the asymmetry of the structure is expressed in the fact that each cell will occupy a unique position in the structure and have a unique potential in the morphogenetic field, which was predicted by many theorists of biology [39, 40, 41, 42]. The unique potential of the cell in the morphogenetic field activates a certain part of DNA and results in the differentiation of embryonic cells which form various tissues and body parts.

The overwhelming majority of living beings have one axis of mirror symmetry, which indicates that the first division of the fertilized cell is symmetrical (Figure A.3), and then ubiquitously asymmetric.

The author wants to draw attention to several indirect evidences in favor of F-division:

Identical (monozygotic) twins are usually born in twos, sometimes threes, very rarely five, but never in four; only 4 is not a member of the Fibonacci sequence.

The ratio of linear dimensions from the crown to the navel and from the navel to the feet in an adult is 1:τ (Figure A.1), in a newborn 1:1, and in a three-month embryo τ:1.

It can also be assumed that the spontaneous uncontrolled cell division in a cancer tumor is a pathological transition from asymmetric division to symmetric division.

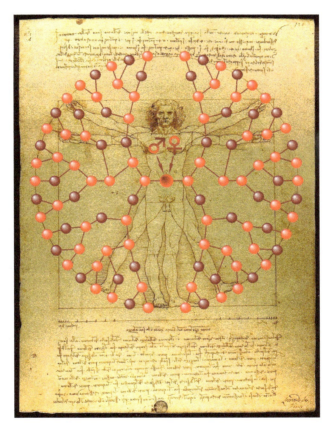

Figure A.3. The F-division with first symmetrical division.

In 2010, while working on [43], the author was surprised to find in Jean [3] a drawing with the F-division scheme, taken from Berdyshev [20]. As the analysis of Berdyshev's article showed, the author was not the first who suggested that cells division process happened according to the "Fibonacci Law". It is surprising that Berdyshev proposed a scheme for asymmetric cell division precisely on the basis of numerical ratios in the inflorescence of sunflowers and other solanaceous cultures and never mentioned the Golden Section or proportions of the human body. In addition, Berdyshev in [20] suggested cell reproduction, and not cell division, takes place. According to Berdyshev when cells reproduce, the mother cell creates a new (daughter) cell in each division cycle, and the daughter misses one division after birth. Well, almost like in the Fibonacci Rabbit problem!

Paradoxically, both articles, discovering and describing asymmetric cell division, were written in Russian.

A.3. THE GOLDEN SECTION IN TECHNIQUE

In the second half of the 20th century, many researchers believed that the use of the Golden Section as the Law of Harmony would give a significant effect in engineering. However, this did not happen. Even manufacturers of televisions and monitors, which originally produced screens with a ratio of length and width close to the Golden Ratio, abandoned the "gold standard" in favor of increasing the width of the screen.

According to the author, the crucial role was played by the fundamental difference in the morphological processes in nature and technology.

In nature, an organism develops from a single cell and exists as an indivisible community of cells, as the continuity of parts in a whole. In contrast to nature, within technology, the final object is created from a set of objects which were created separately. For example, a wheel, which was probably the first technological object. A wheel is not just a round disk, but a disk with a hole in the middle and an axis inserted into it. This is how the main property of technology is manifested — the discontinuity of parts in whole. Indeed, the disk and the axis are made separately, and then connected into a mechanism. The connection of the disk and the axis, especially the improved bearing with lubricant, is much more effective than the kneecap, which has a similar function in a living organism.

The morphology of the living nature is limited by the framework of a unified and indivisible object — a living organism, and the morphology in technology is limited only by the physical properties of the parts which were made independently from each other. Therefore, the direct use of the Golden Section in technique does not bring the expected effect.

A.4. WHY DOES NATURE NEED THE GOLDEN SECTION?

Many years ago, in a private conversation, Professor Ivan Tkachenko told me: your morphological process (author's note — the Fibonacci numbers at the integer level and the Golden Section at the statistical) had a continuation. The square of the Golden Section $\tau^2 = 2.618034...$ is a good approximation of the number $e = 2.718282...$, the difference is only 3.69%. Through the Fibonacci numbers and the Golden Section, nature tends to a ratio of parts in whole, which is close to the number e, and this makes it possible to obtain the maximum energy conversion efficiency.

One should notice that the morphogenetic chain

The Fibonacci numbers → Golden Section → number e

displays a hierarchy of types of numbers in mathematics:

integer → irrational → transcendental

APPENDIX

EXTENSIONS TO THE FIBONACCI SEQUENCE AND THE GOLDEN SECTION

The Fibonacci sequence, like everything related to the Golden Section and recursion, produces a fascinating effect on many researchers and stimulates the search for new numerical relations containing various types of recursions.

In Section II.2, we have already considered the classical Fibonacci sequence and its extension to a recurrent sequence with arbitrary initial terms, also called a generalized recurrent sequence or a Fibonacci sequence with arbitrary initial terms.

B.1. p-NUMBERS FIBONACCI

Let's return to Figure X.1, depicting Pascal's triangle, and drawing another family of oblique lines, shown in green in Figure B.1:

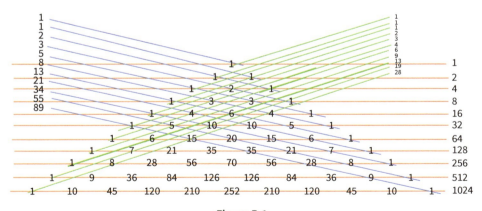

Figure B.1

Green lines pass through binomial coefficients $\binom{n-2i}{2i}$

$$\sum_{i=0}^{n/2}\binom{n-2i}{2i} = F_{2,n}$$

Let's find the sum of these binomial coefficients:

$$\sum_{i=0}^{n/4} \binom{n-2i}{2i} = F_{2,n}$$

Then we get the sequence 1, 1, 1, 2, 3, 4, 6, 9, 13, 19, 28. This sequence is formed by the recurrence formula [23]:

$$F_n = F_{n-1} + F_{n-3}$$

Let's draw inclined lines through the binomial coefficients $\binom{n-3i}{3i}$, then we'll obtain a sequence 1, 1, 1, 1, 2, 3, 4, 5, 7, 10, 14, 19, 26, which is formed by the recurrence formula [23]:

$$F_n = F_{n-1} + F_{n-4}$$

Proceeding from this, Stakhov [23, 25] defined p-Fibonacci numbers as a recurrent sequence, which is formed by the formula

$$F_{p,n} = F_{p,n-1} + F_{p,n-2}$$

as well as p-Golden Sections

$$\lim_{n \to \infty} \frac{F_{p,n}}{F_{p,n-1}} = \tau_p$$

τ_0	τ_1	τ_2	τ_3	τ_4
2	1.618..	1.465..	1.380..	1.324..

B.2. METALLIC MEANS

Another extension of the recursive sequence is the Fibonacci m-numbers or *Metallic Means* by Vera Spinadel [44, 45]. The Fibonacci m-numbers are formed by the recurrence formula:

$$F_{m,n} = mF_{m,n-1} + F_{m,n-2}$$

Metallic Means are calculated by the formula:

$$\lim_{n \to \infty} \frac{F_{m,n}}{F_{m,n-1}} = \frac{\sqrt{4+m^2}+m^2}{2}$$

m			Nick name
0	$\dfrac{\sqrt{4+0^2}+0}{2}$	1	
1	$\dfrac{\sqrt{4+1^2}+1}{2}$	1.618033989	Golden
2	$\dfrac{\sqrt{4+2^2}+2}{2}$	2.414213562	Silver
3	$\dfrac{\sqrt{4+3^2}+3}{2}$	3.302775638	Bronze
4	$\dfrac{\sqrt{4+4^2}+4}{2}$	4.236067978	
5	$\dfrac{\sqrt{4+5^2}+5}{2}$	5.192582404	

B.3. CONTINUOUS FIBONACCI FUNCTIONS

The idea of the possibility of the existence of continuous Fibonacci functions appeared in Stakhov and Tkachenko [46] from the analysis of Binet's formula:

$$F_n = \frac{\tau^n - (-1)^n \tau^{-n}}{\sqrt{5}} \quad (B.1)$$

From (B.1), it can be seen that to replace discrete n with continuous x, it is necessary to replace $(-1)^n$ with a continuous function. Stakhov and Tkachenko [46] selected the replacement of the "discrete" Binet's formula for two functions:

$$F_n = \begin{cases} \dfrac{\tau^{2k+1} + \tau^{-(2k+1)}}{\sqrt{5}}, & \text{if } n = 2k+1 \\ \dfrac{\tau^{2k} - \tau^{-2k}}{\sqrt{5}}, & \text{if } n = 2k \end{cases}$$

From here were identified:

The hyperbolic Fibonacci sine $\quad sF(x) = \dfrac{\tau^{2x} - \tau^{-2x}}{\sqrt{5}}$

The hyperbolic Fibonacci cosine $\quad cF(x) = \dfrac{\tau^{2x+1} + \tau^{-(2x+1)}}{\sqrt{5}}$

The development of hyperbolic Fibonacci functions was continued by Stakhov and Rozin in [47]. Symmetrical Hyperbolic Fibonacci functions (Figure B.2) were defined:

$$\text{Symmetrical Hyperbolic Fibonacci sine} \quad sFs(x) = \frac{\tau^x - \tau^{-x}}{\sqrt{5}}$$

$$\text{Symmetrical Hyperbolic Fibonacci cosine} \quad cFs(x) = \frac{\tau^x + \tau^{-x}}{\sqrt{5}}$$

The Fibonacci numbers are determined identically through the symmetrical hyperbolic Fibonacci functions as the following:

$$F_n = \begin{cases} cFs(n), & \text{for } n = 2k+1 \\ sFs(n), & \text{for } n = 2k \end{cases}$$

The introduced above "symmetrical hyperbolic Fibonacci functions" are connected with the classical hyperbolic functions by the following simple correlations:

$$sFs(x) = \frac{2}{\sqrt{5}} \sinh(\ln(\tau) \cdot x)$$

$$cFs(x) = \frac{2}{\sqrt{5}} \cosh(\ln(\tau) \cdot x)$$

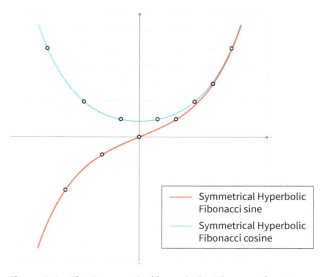

Figure B.2. The Symmetrical hyperbolic Fibonacci functions.

The following correlations that are analogous to the recurrent equation (II.3) for the Fibonacci numbers are valid for the symmetrical hyperbolic Fibonacci functions:

The identities for the Fibonacci numbers	The identities for the symmetrical hyperbolic Fibonacci functions	
$F_{n+2} = F_{n+1} + F_n$	$sFs(x+2) = cFs(x+1) + sFs(x)$	$cFs(x+2) = sFs(x+1) + cFs(x)$
$F_n = (-1)^n F_{-n}$	$sFs(x) = -sFs(-x)$	$cFs(x) = cFs(-x)$
$F_{n+3} + F_n = 2F_{n+2}$	$sFs(x+3) + cFs(x) = 2cFs(x+2)$	$cFs(x+3) + sFs(x) = DHFs(x+2)$
$F_{n+3} - F_n = 2F_{n+1}$	$sFs(x+3) - cFs(x) = DHFs(x+1)$	$cFs(x+3) - sFs(x) = 2cFs(x+1)$
$F_{n+6} - F_n = 4F_{n+3}$	$sFs(x+6) + sFs(x) = 4cFs(x+3)$	$cFs(x+6) + cFs(x) = 4sFs(x+3)$
$F_n^2 - F_{n+1}F_{n-1} = (-1)^{n+1}$	$[sFs(x)]^2 - cFs(x+1) \cdot cFs(x-1) = -1$	$[cFs(x)]^2 - sFs(x+1) \cdot sFs(x-1) = 1$
$F_{2n+1} = F_{n+1}^2 + F_n^2$	$cFs(2x+1) = [cFs(n+1)]^2 + [cFs(x)]^2$	$cFs(2x+1) = [sFs(n+1)]^2 + [sFs(x)]^2$

The next continuous function defined by Stakhov and Rozin [48] is the quasi-sine Fibonacci function (Figure B.3), which is passing through the Fibonacci numbers:

$$FF(x) = \frac{\tau^x - \cos(\pi x)\tau^{-x}}{\sqrt{5}}$$

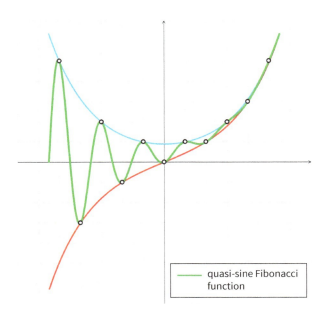

Figure B.3. The quasi-sine Fibonacci function.

The identities for the Fibonacci numbers	The identities for the quasi-sine Fibonacci function
$F_{n+2} = F_{n+1} + F_n$	$FF(x+2) = FF(x+1) + FF(x)$
$F_{n+3} + F_n = 2F_{n+2}$	$FF(x+3) + FF(x) = 2FF(x+2)$
$F_{n+3} - F_n = 2F_{n+1}$	$FF(x+3) - FF(x) = 2FF(x+1)$
$F_{n+6} - F_n = 4F_{n+3}$	$FF(x+6) + FF(x) = 4FF(x+3)$
$F_n^2 - F_{n+1} F_{n-1} = (-1)^{n+1}$	$[FF(x)]^2 - FF(x+1) \cdot FF(x-1) = -1$
$F_{2n+1} = F_{n+1}^2 + F_n^2$	$FF(2x+1) = [FF(x+1)]^2 + [FF(x)]^2$

Turning to the complex 3D space, Stakhov and Rozin [48] determined the complex function Three-dimensional Fibonacci spiral (Figure B.4):

$$CFF(x) = \frac{\tau^x - \cos(\pi x)\tau^{-x}}{\sqrt{5}} + i\frac{\sin(\pi x)\tau^{-x}}{\sqrt{5}},$$

for which the recurrence equation will also be executed (II.3):

$$CFF(x+2) = CFF(x+1) + CFF(x)$$

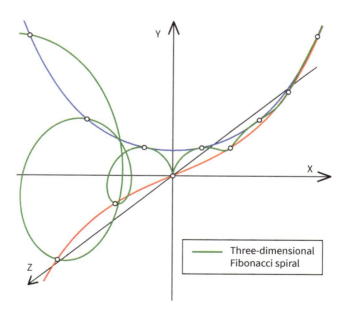

Figure B.4. Three-dimensional Fibonacci spiral.

Three-dimensional Fibonacci spiral pierces the XY plane at points (n, F_n) and its projection on the XY plane is the quasi-sine Fibonacci function:

$$Re(CFF(x)) = \frac{\tau^x - \cos(\pi x)\tau^{-x}}{\sqrt{5}}$$

$$Im(CFF(x)) = \frac{\sin(\pi x)\tau^{-x}}{\sqrt{5}}$$

From these formulas, Stakhov and Rozin [48] defined the system

$$\begin{cases} y(x) - \dfrac{\tau^x}{\sqrt{5}} = -\dfrac{\cos(\pi x)\tau^{-x}}{\sqrt{5}} \\ z(x) = \dfrac{\sin(\pi x)\tau^{-x}}{\sqrt{5}} \end{cases},$$

which defines a 3D surface, which they called the Golden Shofar [48] (Figure B.5):

$$\left(y - \frac{\tau^x}{\sqrt{5}}\right)^2 - z^2 = \left(\frac{\tau^{-x}}{\sqrt{5}}\right)^2$$

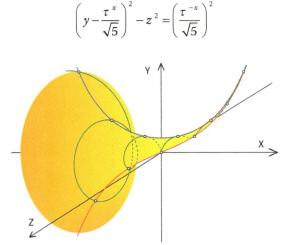

Figure B.5. The Golden Shofar.

The number of extensions of the Fibonacci sequence is constantly increasing; here are some of them:

- Binet's formulas and Continuous Functions for p-Fibonacci Numbers [49, 50, 51]
- the Fibonacci Q–matrix [52]

- Fibonacci λ-numbers and Fibonacci λ-functions [53]
- k–Fibonacci numbers and k–Fibonacci hyperbolic functions [54]

which, along with those described above, Stakhov called *The Mathematics of Harmony* [24].

B.4. BODNAR'S "GOLDEN" HYPERBOLIC FUNCTIONS

Ukrainian architect Bodnar [55] investigated the hyperbolic rotation on the planar representation of the cylindrical Fibonacci phyllotaxis, at which the index of visible parastichies increases from (F_{n-1}, F_n) to (F_n, F_{n+1}). As a result of this analysis, Bodnar received functions which he called "golden hyperbolic functions":

The "golden" hyperbolic sine $\quad Gsh(x) = \dfrac{\tau^x - \tau^{-x}}{2}$

The "golden" hyperbolic cosine $\quad Gch(x) = \dfrac{\tau^x + \tau^{-x}}{2}$

Let me remind you that the hyperbolic rotation [56] is the transformation of a Cartesian coordinate system, in which, at the same time, compression occurs k times on one of the axes and stretching k times the other. With such a transformation, each point moves along a hyperbola (hence the name), the parallel lines stay parallel, and the areas of the figures remain constant. For example, a hyperbolic rotation transforms a square into a parallelepiped with the same area.

Hyperbolic rotation by τ means stretching a Cartesian coordinate system along the OX axis by τ times and compression along the OY axis by τ times. If we make a hyperbolic turn on τ over the planar representation of the cylindrical Fibonacci phyllotaxis with the index (F_n, F_{n+1}), then it is transformed into a lattice with the index (F_{n+1}, F_{n+2}).

In terms of the DH-Model, a hyperbolic rotation by τ times increases the cylinder radius in τ times and reduces the genetic helix pitch in τ times, that is, the ratio of the radius to the helix pitch increases τ^2 times. As it was proven in Section VII.1 of this study, "the phyllotaxis rises" occurs at any change in the ratio of the radius to the pitch of the helix more than τ^2 times (see Video VII.1).

Thus, a hyperbolic rotation on τ is a particular case of the transformation of the planar representation of the cylindrical Fibonacci phyllotaxis, leading to "the phyllotaxis rises". However, in the general case, the stretching and/or compression of the planar representation of the cylindrical Fibonacci phyllotaxis, which leads to "the phyllotaxis rises", does not produce such beautiful formulas as "golden hyperbolic functions."

For the sake of justice, the author wants to note that a detailed examination of the method by which Professor Bodnar received the "golden" hyperbolic functions allowed the author to see many lacunae in mathematical phyllotaxis and gave impetus to the beginning of these studies.

B.5. THE GOLDEN SINE

The author called the sine-like function *the Golden sine*, the points of intersection of which with the axis **OX** form segments, the ratio of the lengths of which is equal to the Golden Section:

$$Gsin(x) = \sin(\pi \log_\tau(x))$$

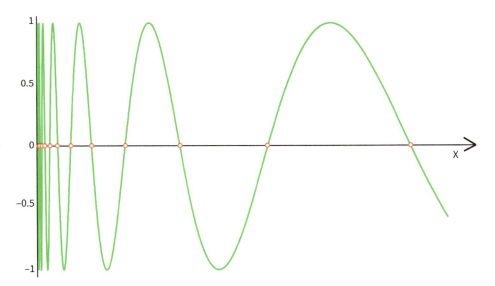

Figure B.6. The Golden sine function.

The graph of the Golden sine function (Figure B.6) crosses the OX axis in points $x = \tau^n$, where $n \in \mathbb{Z}$.

If we compare the lengths of two segments, which are formed by points $(\tau^{n-2}, 0)$, $(\tau^{n-1}, 0)$, $(\tau^n, 0)$, then their ratio is equal to the Golden Section:

$$\frac{\tau^n - \tau^{n-1}}{\tau^{n-1} - \tau^{n-2}} = \tau$$

When stretching (or compressing) the graph of the Golden sine function τ^2 times along the OX axis, the stretched graph will coincide with the original:

$$Gsin(x \cdot \tau^2) = \sin(\pi \log_\tau(x \cdot \tau^2)) = \sin(\pi \log_\tau(x) + 2\pi) = \sin(\pi \log_\tau(x)) = Gsin(x)$$

Let's multiply the Golden sine function by $\frac{1}{x}$, then we'll get *the Truncated Golden sine* function:

$$TGsin(x) = \frac{1}{x}\sin(\pi\log_\tau(x))$$

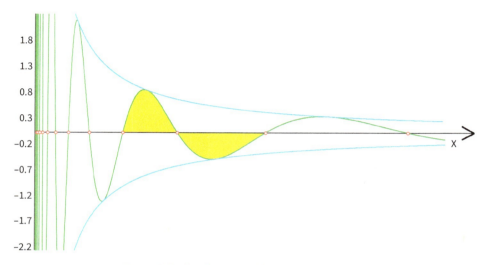

Figure B.7. The Truncated Golden sine function.

This function (Figure B.7) is interesting because the areas bounded by the half-period graph and the axis *OX* are equal:

$$\int_{\tau^{n-1}}^{\tau^n} |TGsin(x)|\,dx = \int_{\tau^{n-1}}^{\tau^n} \left|\frac{1}{x}\sin(\pi\log_\tau(x)x)\right|dx$$

Let's replace $z = \log_\tau(x)$, when:

$$\frac{dz}{dx} = \frac{1}{x\ln(\tau)}$$

$$\int_{\tau^{n-1}}^{\tau^n} \left|\frac{1}{x}\sin(\pi\log_\tau(x)x)\right|dx = \int_{n-1}^{n} \ln(\tau)|\sin(\pi z)|\,dz = \frac{\ln(\tau)}{\pi}|\cos(\pi z)|\Big|_{n-1}^{n} =$$

$$= \frac{\ln(\tau)}{\pi}|\cos(\pi n) - \cos(\pi(n-1))| = \frac{2\cdot\ln(\tau)}{\pi} = const$$

Extensions to the Fibonacci Sequence and the Golden Section **153**

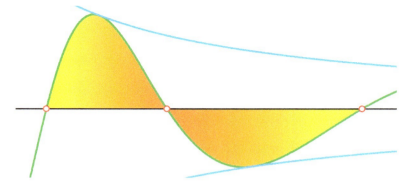

Figure B.8. Graph of one period of the Truncated Golden sine function.

Figure B.8 could be an explanation of the "golden heartbeat" by Tsvetkov [57]. Short half period – systole, long – diastole.

APPENDIX

THE FIBONACCI NUMBERS IN CYBERNETICS

At the beginning of the computer era, when calculations were made on tube or transistor devices, there was an acute problem of reliability of calculations, because the first computing systems were used to calculate the trajectory and control of ballistic missiles. Therefore, in the 60s–70s of the last century, research was in demand to improve the reliability of computers at the hardware level. One direction was the use of redundant binary codes with the ability to detect and/or correct errors.

Bergman was the first who proposed to use the Golden Section for coding [58]. However, this idea was ignored by mathematicians, perhaps because Bergman was only 14 years old. Bergman himself subsequently did not see the real application of the Golden Section codes. Even on his Wikipedia [59] this article is not mentioned.

The Golden Section codes were rediscovered by Vitenko and Stakhov [23] in 1970.

Modern computers use the classical binary number system, in which the weights of binary digits are equal to the power of the number two; each number can be represented as:

$$r = \sum_{i=0}^{n} a_i 2^i$$

where a_i may be 0 or 1. Moreover, there is a one-to-one correspondence between the number and its code. For example, the number 11 will have a binary code 01011.

The idea of the Golden Section codes and the Fibonacci codes is that the weights of binary digits are equal to the degree of the Golden Section base or the Fibonacci numbers. For example, in the Fibonacci codes, each number can be represented as:

$$r = \sum_{i=0}^{n} a_i F_i$$

where $F_0 = 1$, $F_0 = 1$, $F_{i+2} = F_{i+1} + F_i$. Unlike classic binary codes, a single number can have multiple images. For example, the number 11 will be displayed in the Fibonacci code as 0101000, 0100110 or 0011101. This allows you to separate the images of codes into the correct and wrong (allowed and prohibited). Stakhov [24] chose the correct images of codes with the minimum number of units and called them the minimum form. Images of codes that have a minimum number of units are considered forbidden, which allows detecting a certain percentage of errors. During the 1970s–1980s, under the leadership of Stakhov, an adder (ALU) was developed, and then a processor operating in the Fibonacci code.

However, the emergence of universal high-speed and relatively cheap microprocessors made it possible to transfer the solution to the problem of reliability of data processing and transmission to the software area, and to use majority architecture in control systems. Classic binary codding has supplanted all others from commercial use forever.

According to the author, the most interesting was the development of an analog-to-digital converter (ADC) in the Golden Section codes, in which the weights of the input bits differed by 1.618 times. That is, if one weighing discharge fails, it is automatically replaced by two younger ones (formula (II.2) without loss of measurement accuracy. Thus, the fault tolerance of the entire converter was significantly increased.

In the early 2000s, the author corresponded with Dr Poulton, the leading Agilent Laboratories specialist in Silicon Valley. After my story about the ADC with weights of digits equal to τ^n, my correspondence with Dr. Poulton can be represented in a dialogue:

- Oh, thank you, young man, I learned a lot of funny things, something I had never suspected. We have already solved the problem of failure of one discharge and replacing it with two younger ones, the solution of which you proposed.
- But how?
- We use, for weighing, a 12-bit converter with weights of digits $G = 1.6$ (*author* – each following digit is 1.6 times greater than the previous one), and then convert it into an 8-bit binary code, which is the output of the ADC chip.
- Where can I read about it?
- See page 10 from [60].
- How did you get that the weights of the discharges should be in the power of the Golden Section?
- Oh, we didn't even know that 1.6 is such a famous constant. The number 1.6 was obtained experimentally.

APPENDIX

THE ANGLE OF THE TANGENT TO THE SPIRAL

As it is known from analytic geometry, if we let draw a tangent to any point of the graph of a continuous function, then the tangent of the angle between this tangent and the axis *OX* will be equal to the derivative of this continuous function at the point of tangency.

Figure D.1. A tangent line to the helix shows a helix (red curve), a tangent to the helix (blue straight line), a radius-vector passing through the tangency point (magenta line), an *OX* axis (black straight line). Also, δ is the angle of inclination of the tangent to the *OX* axis, γ is the angle between the tangent and the radius vector, θ is the angle of rotation, the radius of the vector from 0 to the tangency point.

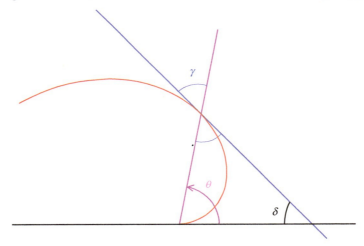

Figure D.1. A tangent line to the helix.

Let's find the tangent function of angle δ as a derivative of the helix (II.15) at the tangency point:

$$\tan(\delta) = \frac{dy}{dx} = \frac{\frac{dy_\theta}{d\theta}}{\frac{dx_\theta}{d\theta}} = \frac{\frac{d}{d\theta}(E(\theta)\cdot\sin(k\theta))}{\frac{d}{d\theta}(E(\theta)\cdot\cos(k\theta))} = \frac{(E(\theta))'\cdot\sin(k\theta)+k\cdot E(\theta)\cdot\cos(k\theta)}{(E(\theta))'\cdot\cos(k\theta)-k\cdot E(\theta)\cdot\sin(k\theta)}$$

If we divide the numerator and denominator by $(E(\theta))' \cdot \cos(k\theta)$, then:

$$\tan(\delta) = \frac{\tan(k\theta) + \dfrac{k \cdot E(\theta)}{(E(\theta))'}}{1 - \tan(k\theta) \cdot \dfrac{k \cdot E(\theta)}{(E(\theta))'}}$$

The last formula is very similar to the tangent function of the sum of the angles:

$$\tan(\delta) = \tan(\alpha + \gamma) = \frac{\tan(\alpha) + \tan(\gamma)}{1 - \tan(\alpha) \cdot \tan(\gamma)}$$

Let $\alpha = k\theta$, then:

$$\tan(\gamma) = \frac{k \cdot E(\theta)}{(E(\theta))'}$$

From Figure D.1 follows that $0 \leq \gamma < \dfrac{\pi}{2}$, accordingly:

$$\gamma = \arctan\left(\left|\frac{k \cdot E(\theta)}{(E(\theta))'}\right|\right) \qquad (D.1)$$

For the Archimedean spiral $E(\theta) = \theta$, then $(E(\theta))' = 1$, let's substitute it in (D.1):

$$\gamma = \arctan\left(\left|\frac{k \cdot \theta}{1}\right|\right) = \arctan\left(|k \cdot \theta|\right)$$

For power spiral $E(\theta) = \theta^\nu$, then $(E(\theta))' = \nu\theta^{\nu-1}$, let's substitute it in (D.1):

$$\gamma = \arctan\left(\left|\frac{k \cdot \theta^\nu}{\nu\theta^{\nu-1}}\right|\right) = \arctan\left(\left|\frac{k}{\nu}\theta\right|\right)$$

For logarithmic spiral $E(\theta) = e^{b\theta}$, then $(E(\theta))' = be^{b\theta}$, let's substitute it in (D.1):

$$\gamma = \arctan\left(\left|\frac{k \cdot e^{b\theta}}{be^{b\theta}}\right|\right) = \arctan\left(\left|\frac{k}{b}\right|\right) = \text{constant}$$

From the last formula, the remarkable property of a logarithmic spiral follows: the angle between the tangent and the radius vector is the same for all points of the spiral.

APPENDIX

SOLVE A LINEAR DIOPHANTINE EQUATION

This example is based on video [61]. It is necessary to find integer solutions of the equation

$$Ax - By = 1, \text{ where } A, B \in N, A > B, \gcd(A, B) = 1.$$

Let's use the decomposition of the number to find a solution $A = B \cdot q + r$, 1 which we are going to write in the form $A = B(q) + r$.

For example, let's find the solution for the equation $47x - 30y = 1$ or $47x + 30(-y) = 1$.

$$47 = 30(1) + 17$$
$$\downarrow \quad \downarrow$$
$$30 = 17(1) + 13$$
$$\downarrow \quad \downarrow$$
$$17 = 13(1) + 4$$
$$\downarrow \quad \downarrow$$
$$13 = 4(3) + 1$$

After the remainder of the division is equal to unity ($r = 1$), we rewrite the resulting equations in the form of differences:

$$17 = \mathbf{47}(1) + \mathbf{30}(-1)$$
$$13 = \mathbf{30}(1) + \mathbf{17}(-1)$$
$$4 = \mathbf{17}(1) + \mathbf{13}(-1)$$
$$1 = \mathbf{13}(1) + \mathbf{4}(-3)$$

Let's substitute the value of **4** from the penultimate equation into the last equation:

$$1 = \mathbf{13}(1) + \mathbf{4}(-3) = \mathbf{13}(1) + (-3)(\mathbf{17}(1) + \mathbf{13}(-1)) = \mathbf{13}(4) + \mathbf{17}(-3)$$

Let's substitute the value of **13** into the last equation:

$$1 = \mathbf{13}(4) + \mathbf{17}(-3) = (4)\,(\mathbf{30}(1) + \mathbf{17}(-1)) + \mathbf{17}(-3) = \mathbf{30}(4) + \mathbf{17}(-7)$$

Let's substitute the value of **17** into the last equation:

$$1 = \mathbf{30}(4) + \mathbf{17}(-7) = \mathbf{30}(4) + (-7)\,(\mathbf{47}(1) + \mathbf{30}(-1)) = \mathbf{47}(-7) + \mathbf{30}(11)$$

into the last equation $x = -7$ and $y = 11$.

APPENDIX

TABLE OF THE DEPENDENCE OF THE DIVERGENCE COEFFICIENT AND ENCYCLIC NUMBERS ON THE INITIAL TERMS OF THE GENERATING RECURRENT SEQUENCE

G_1	G_2	A_1	A_2	B_1	B_2	β	$360° \cdot \beta$
1	2	0	1	1	1	0.38197	137.50776
1	3	0	1	1	2	0.27639	99.50155
1	4	0	1	1	3	0.21654	77.95525
1	5	0	1	1	4	0.17800	64.07936
1	6	0	1	1	5	0.15110	54.39682
1	7	0	1	1	6	0.13127	47.25629
1	8	0	1	1	7	0.11604	41.77287
1	9	0	1	1	8	0.10397	37.42969
1	10	0	1	1	9	0.09418	33.90458
1	11	0	1	1	10	0.08607	30.98631
1	12	0	1	1	11	0.07925	28.53059
1	13	0	1	1	12	0.07343	26.43553
1	14	0	1	1	13	0.06841	24.62711
1	15	0	1	1	14	0.06403	23.05028
1	16	0	1	1	15	0.06018	21.66321
1	17	0	1	1	16	0.05676	20.43361
1	18	0	1	1	17	0.05371	19.33609
1	19	0	1	1	18	0.05097	18.35046
1	20	0	1	1	19	0.04850	17.46044
1	21	0	1	1	20	0.04626	16.65276
1	22	0	1	1	21	0.04421	15.91650
2	5	1	2	1	3	0.41982	151.13566
2	7	1	3	1	4	0.43929	158.14491
2	9	1	4	1	5	0.45115	162.41512

G_1	G_2	A_1	A_2	B_1	B_2	β	$360° \cdot \beta$
2	11	1	5	1	6	0.45914	165.28939
2	13	1	6	1	7	0.46488	167.35606
2	15	1	7	1	8	0.46920	168.91357
2	17	1	8	1	9	0.47258	170.12945
2	19	1	9	1	10	0.47529	171.10499
2	21	1	10	1	11	0.47751	171.90504
2	23	1	11	1	12	0.47937	172.57305
2	25	1	12	1	13	0.48094	173.13922
2	27	1	13	1	14	0.48229	173.62517
2	29	1	14	1	15	0.48346	174.04684
3	7	1	2	2	5	0.29569	106.44696
3	8	1	3	2	5	0.36716	132.17767
3	10	1	3	2	7	0.30521	109.87692
3	11	1	4	2	7	0.35927	129.33554
3	13	1	4	2	9	0.31089	111.92142
3	14	1	5	2	9	0.35436	127.56902
3	16	1	5	2	11	0.31466	113.27886
3	17	1	6	2	11	0.35101	126.36466
3	19	1	6	2	13	0.31735	114.24574
3	20	1	7	2	13	0.34859	125.49096
3	22	1	7	2	15	0.31936	114.96942
4	9	1	2	3	7	0.22821	82.15490
4	11	1	3	3	8	0.26856	96.68046
4	13	1	3	3	10	0.23384	84.18309
4	15	1	4	3	11	0.26431	95.15106
4	17	1	4	3	13	0.23716	85.37801
4	19	1	5	3	14	0.26164	94.19148
4	21	1	5	3	16	0.23935	86.16567
4	23	1	6	3	17	0.25981	93.53327
4	25	1	6	3	19	0.24090	86.72395
4	27	1	7	3	20	0.25848	93.05373
5	11	1	2	4	9	0.18581	66.89005
5	12	2	5	3	7	0.41325	148.77132
5	13	2	5	3	8	0.38757	139.52522
5	14	1	3	4	11	0.21170	76.21295
5	16	1	3	4	13	0.18952	68.22843
5	17	2	7	3	10	0.40996	147.58384

Table of the Dependence of the Divergence Coefficient

G_1	G_2	A_1	A_2	B_1	B_2	β	$360° \cdot \beta$
5	18	2	7	3	11	0.39052	140.58609
5	19	1	4	4	15	0.20905	75.25937
5	21	1	4	4	17	0.19170	69.01123
5	22	2	9	3	13	0.40797	146.86965
5	23	2	9	3	14	0.39233	141.24034
5	24	1	5	4	19	0.20738	74.65779
6	13	1	2	5	11	0.15669	56.40895
6	17	1	3	5	14	0.17472	62.89740
6	19	1	3	5	16	0.15933	57.35778
6	23	1	4	5	19	0.17291	62.24650
6	25	1	4	5	21	0.16086	57.91001
6	29	1	5	5	24	0.17176	61.83440
6	31	1	5	5	26	0.16186	58.27130
6	35	1	5	5	30	0.14514	52.24970
6	37	1	6	5	31	0.16257	58.52610
6	41	1	7	5	34	0.17039	61.34204
6	43	1	7	5	36	0.16310	58.71543
6	47	1	8	5	39	0.16995	61.18324
7	15	1	2	6	13	0.13547	48.76750
7	16	3	7	4	9	0.43560	156.81587
7	17	2	5	5	12	0.29241	105.26866
7	18	2	5	5	13	0.27932	100.55364
7	19	3	8	4	11	0.42245	152.08096
7	20	1	3	6	17	0.14873	53.54269
7	22	1	3	6	19	0.13743	49.47506
7	23	3	10	4	13	0.43380	156.16774
7	24	2	7	5	17	0.29076	104.67272
7	25	2	7	5	18	0.28084	101.10347
7	26	3	11	4	15	0.42386	152.58987
7	27	1	4	6	23	0.14742	53.07028
7	29	1	4	6	25	0.13857	49.88539
7	30	3	13	4	17	0.43273	155.78394
7	31	2	9	5	22	0.28976	104.31296
7	32	2	9	5	23	0.28178	101.44140
7	33	3	14	4	19	0.42474	152.90790
7	34	1	5	6	29	0.14658	52.77043

REFERENCES

1. L. Bravais and A. Bravais, "Essai sur la disposition generate des feuilles," *Ann. Sci. Nat. Bot.*, vol. 12, pp. 5–14, 65–77, 1839.
2. I. Adler, D. Barabe and R. V. Jean, "A History of the Study of Phyllotaxis," *Annals of Botany*, vol. 80, pp. 231–244, 1997.
3. R. Jean, Phyllotaxis. A Systemic Study in Plant Morphogenesis, Cambridge University Press, 1994.
4. A. H. Church, On the Interpretation of Phenomena of Phyllotaxis, New York: Hafner Pub. Co., 1920 (Reprinted in 1968).
5. A. M. Turing, "Morphogen theory of phyllotaxis," in *Morphogenesis*, vol. 3, Elsevier, 1992.
6. A. M. Turing, "Morphogen theory of phyllotaxis. Part I. Geometrical and descriptive phyllotaxis," [Online]. Available: http://www.turingarchive.org/browse.php/C/8.
7. L. S. Levitov, "Phyllotaxis of flux lattices in layered superconductors," *Physical Rewiew Letters*, vol. 66, no. 2, pp. 224–7, 1991.
8. I. Adler, Solving the Riddle of Phyllotaxis: Why the Fibonacci Numbers and the Golden Ratio Occur on Plants, World Scientific, 2012.
9. H. S. M. Coxeter, "The Golden Section and Phyllotaxis," in *Introduction to Geometry*, Wiley, 1969.
10. S. Vajda, Fibonacci and Lucas Numbers, and the Golden Section: Theory and Applications, Halsted Press, 1989.
11. D. Lina, "116 Photos Of Geometrical Plants For Symmetry Lovers," [Online]. Available: https://www.boredpanda.com/geometry-symmetry-plants-nature/?utm_source=google&utm_medium=organic&utm_campaign=organic.
12. T. M. E. Foundation, "M.C. Escher Collection," THE M.C. ESCHER COMPANY B.V., [Online]. Available: https://mcescher.com/gallery/most-popular/.
13. "Bernard Pras' anamorphic portrait of Ferdinand Cheval," [Online]. Available: https://thekidshouldseethis.com/post/bernard-pras-anamorphic-portrait-of-ferdinand-cheval.

14. A. M. Turing, "The Chemical Basis of Morphogenesis," *Philosophical Transactions of the Royal Society of London. Series B, Biological Sciences*, vol. 237, no. 641, pp. 37–72, 14 Aug 1952.
15. R. V. Jean, "Number–theoretic properties of two–dimensional lattices," *Journal of Number Theory*, vol. 29, pp. 206–223, 1988.
16. B. Mandelbrot, The fractal geometry of nature, Freeman, 1977.
17. J. Kraft and L. C. Washington, An Introduction to Number Theory with Cryptography, CRC Press, Taylor & Francis Group, 2018.
18. D. Weise, "Phyllotaxis is Not Logarithmic //Interdisciplinary Conference Symmetry of Forms and Structures Symmetry," 2009. [Online]. Available: https://cloud.mail.ru/public/KM6H/hUomqdy84.
19. A. H. Church, On the Relation of Phyllotaxis to Mechanical Law, London: Williams and Norgate, 1904.
20. A. P. Berdyshe, "On some mathematical regularities of biological processes," *Zhurnal Obshchey Biologii*, vol. 33, pp. 631–8, 1972.
21. L. F. Hernandez and J. H. Palmer, "Regeneration of the Sunflower Capitulum after Cylindrical Wounding of the Receptacle," *American Journal of Botany*, vol. 79, no. 8, pp. 1253–61, 1998.
22. C. Golé and J. Dumais, "Collaborative Research: New Methods in Phyllotaxis," [Online]. Available: http://www.math.smith.edu/~cgole/PHYLLOH/ProjectDescriptionFinal05.pdf.
23. A. P. Stakhov, Introduction into Algorithmic Measurement Theory, Moscow: Soviet Radio, 1977.
24. A. Stakhov, The Mathematics of Harmony: From Euclid to Contemporary Mathematics and Computer Science, World Scientific, 2009.
25. A. P. Stakhov, "The Golden Section in the Measurement Theory," *Computers & Mathematics with Applications*, vol. 17, pp. 4–6, 1989.
26. D. E. Knuth, The Art of Computer Programming, Addison–Wesley Publishing Co, 1969.
27. I. Kant, Metaphysical Foundations of Natural Science, Cambridge University Press, 2004.
28. A. Zeising, *Neue Lehre von den Proportionen des menschlichen Korpers*, Leipzig: Rudolph Weigel, 1854.
29. G. Doczi, The power of limits: proportional harmonies in nature, art, and architecture, Shambhala Publications, 1981.
30. E. Moessel, Die Proportion in Antike und Mittelalter, C.H. Beck, 1926.
31. L. Corbusier, The Modulor: A Harmonious Measure to the Human Scale Universally applicable to Architecture and Mechanics, Harvard University Press, 1954.
32. M. Ghika, The Geometry of Art and Life, New York: Dover, 1977.

33. J. Hambidge, Dynamic symmetry: the Greek vase, Yale university press, 1920.
34. N. A. Solyanichenko and B. N. Rozin, "The mystery of the golden section," in *Unconventional scientific ideas about nature and its phenomena: FENID-90*, 1991.
35. N. A. Solyanichenko and B. N. Rozin, *About the reasons for the presence of the golden ratio in the forms of wildlife*, Kiev: UKR NTI, 1993.
36. B. Rozin, *The golden section is morphology law of living matter*, The Library of Congress. Copyright catalog TXu001103728, 2003.
37. G. T. Fechner, *Vorschule Der Asthetik*, Bod Third Party Titles, 2013.
38. J. Hambidge, Dynamic Symmetry in Composition As Used By the Artists, Yale University Press, 1948.
39. A. G. Gurvich, Die mitogenetische Strahlung: ihre physikalisch-chemischen Grundlagen und ihre Anwendung in Biologie und Medizin (Mitogenetic radiation: its physicochemical basis and its application in biology and medicine), G. Fischer, 1959.
40. L. V. Beloussov, V. L. Voeikov and V. S. Martynyuk, Biophotonics and Coherent Systems in Biology, Springer Science & Business Media, 2007.
41. E. M. De Robertis, E. A. Morita and K. W. Cho, "Gradient fields and homeobox genes," *Development*, vol. 112, pp. 669–678, 1991.
42. B. N. Belintsev, Physical Fundamentals of Biological Formacreation, Moscow: Nauka, 1991.
43. B. Rozin, "Phyllotaxis Model Analysis," 2011 [in Russian]. [Online]. Available: http://www.trinitas.ru/rus/doc/0232/013a/2101-rzn.pdf.
44. V. W. d. Spinadel, From the Golden Mean to Chaos, Nueva Libreria, 1998.
45. V. W. d. Spinadel, "The family of Metallic Means," *Visual Mathematics*, vol. 11, no. 3, 1999.
46. A. P. Stakhov and I. S. Tkachenko, "Hyperbolic Fibonacci trigonometry," *Doklady Akademii Nauk Ukrainy*, vol. 7, pp. 9–14, 1993.
47. A. Stakhov and B. Rozin, "On a new class of hyperbolic function," *Chaos, Solitons & Fractals*, vol. 23, no. 2, pp. 379–89, 2005.
48. A. Stakhov and B. Rozin, "The Golden Shofar," *Chaos, Solitons & Fractals*, vol. 26, no. 3, pp. 677–84, 2005.
49. A. Stakhov and B. Rozin, "Theory of Binet formulas for Fibonacci and Lucas p-numbers," *Chaos, Solitons & Fractals*, vol. 27, no. 5, pp. 1162–77, 2005.
50. A. Stakhov and B. Rozin, "The continuous functions for the Fibonacci and Lucas p-numbers," *Chaos, Solitons & Fractals*, vol. 28, no. 4, pp. 1014–25, 2006.
51. E. Kilic, "The Binet formula, sums and representations of generalized Fibonacci p-numbers," *Chaos, Solitons & Fractals*, vol. 29, no. 3, pp. 701–711, 2008.
52. A. P. Stakhov, "A generalization of the Fibonacci Q-matrix," *Reports of the National Academy of sciences of Ukraine*, vol. 9, pp. 46–9, 1999.

53. M. J. Gazale, Gnomon: From Pharaohs to Fractals, Princeton NJ: Princeton University Press, 1999.
54. S. Falcon and A. Plaza, "The k–Fibonacci hyperbolic functions," *Chaos Solitons Fractals*, vol. 38, no. 2, pp. 409–420, 2008.
55. O. Bodnar, Golden Section and Non–Euclidean Geometry in Science and Art, Lviv: Ukrainian technologies, 2005.
56. V. G. Shervatov et al, Hyperbolic Functions: with Configuration Theorems and Equivalent and Equidecomposable Figures, Dover Publications, 2007.
57. V. D. Tsvetkov, Heart, golden ratio and symmetry, Institute of Theoretical and Experimental Biophysics RAS, 1997 [in Russian].
58. G. Bergman, "A number system with an irrational base," *Mathematics Magazine*, vol. 31, pp. 98–119, 1957.
59. "George Bergman," Wikipedia, the free encyclopedia, [Online]. Available: https://en.wikipedia.org/wiki/George_Bergman.
60. Poulton, Ken and others, "A 20 GSa/s 8b ADC with a 1 MB Memory," Agilent Laboratories, Palo Alto, California, 2003. [Online]. Available: http://poulton.net/papers.public/2003isscc_18_1_pg_slides.pdf.
61. Socratica, "Number Theory: Diophantine Equation: ax+by=gcd(a,b)," [Online]. Available: https://www.youtube.com/watch?v=FjliV5u2IVw.
62. J. W. v. Goethe, The Metamorphosis of Plants, MIT, 2009.

INDEX

A
Angular group, 107, 125, 127, 132
Artificial Intelligence, 126, 127, 128, 132

B
Binet's formula, 6, 32, 46, 50, 59, 83, 91, 105, 145, 149
border of the n-th diapason, 36, 43, 85
Bravais-Bravais theorem, 24, 25, 29, 45, 50, 60, 67

C
cell division, 139
　asymmetric cell division, 139, 140

D
DH-Model, 24, 27, 57, 77, 79, 80, 81, 82, 84, 86, 87, 89, 92, 97, 104, 105, 106, 108, 109, 118, 127, 129, 131, 132, 133, 150
Diophantine equation, 51, 67, 159
divergence coefficient, 23, 25, 28, 29, 30, 31, 45, 46, 50, 51, 52, 59, 66, 67, 68, 69, 74, 79, 80, 82, 93, 97, 98, 99, 100, 101, 102, 105, 106, 107, 108, 131, 132

E
Edge function, 9, 10, 21, 69, 70, 74, 106, 107, 108, 128, 132
encyclic numbers, 25, 57, 59, 68, 106

F
family of parastichies, 14
family of threads, 33, 34, 35, 37, 38, 41, 84, 85, 131
F-division of cells, 138
Fibonacci code, 155
Fibonacci numbers, 1, 2, 5, 6, 8, 14, 16, 17, 18, 32, 48, 59, 68, 92, 115, 116, 117, 118, 136, 138, 139, 140, 141, 143, 144, 146, 147, 148, 149, 150, 155
fractal, 5, 44
fundamental theorem of phyllotaxis, 25, 58, 59, 67, 106, 126
　converse of the fundamental theorem of phyllotaxis, 57, 59, 66, 131

G
Golden Ratio, 3, 4, 5, 8, 135, 140
Golden Section, 1, 3, 6, 135, 136, 138, 140, 141, 143, 151, 155, 156
Golden Shofar, 149
Golden sine, 151, 152, 153

H
helix, 10, 11, 15, 24, 79, 80, 81, 84, 91, 97, 98, 102, 108, 150, 157
hydraulic model, 2, 119, 120, 121, 122, 132, 133
hyperbolic Fibonacci functions, 146
　symmetrical hyperbolic Fibonacci functions, 146, 147

I
index of parastichies.
See parastichy index

L
linear group, 107, 127, 128, 132

M
minimum distance between nearby the nodes, 19, 28, 32, 82, 105
morphogenesis, 19, 20, 106, 108, 109, 110, 114, 115, 122, 126, 132, 133, 135

N
network project "Open Phyllotaxis", 2, 128, 129, 132

O
opposed parastichy pair, 14, 40, 44, 48, 54, 57, 64, 66, 68, 73, 89, 106, 107, 125, 128, 132

P
parastichy index, 14, 25, 38, 44, 57, 66, 85, 89, 107, 109, 125, 127, 128, 132, 150
Pascal's triangle, 116, 117, 143
pattern recognition, 126, 127
phyllotactic lattice.
See phyllotactic pattern
phyllotactic pattern, 14, 15, 18, 19, 21, 22, 23, 24, 25, 27, 28, 29, 31, 33, 34, 37, 38, 39, 40, 42, 44, 45, 46, 47, 48, 50, 51, 52, 53, 54, 55, 57, 59, 60, 61, 62, 64, 65, 66, 67, 68, 69, 71, 72, 76, 77, 78, 80, 84, 86, 88, 89, 90, 93, 94, 96, 97, 98, 99, 100, 101, 102, 103, 104, 105, 106, 107, 108, 110, 111, 112, 113, 118, 119, 120, 121, 122, 125, 127, 128, 13, 23, 133, 100, 103
phyllotaxis, 1, 2, 10, 11, 13, 15, 16, 17, 18, 19, 20, 21, 22, 23, 24, 27, 35, 36, 60, 66, 69, 70, 77, 78, 81, 91, 92, 97, 99, 100, 101, 102, 103, 104, 105, 106, 108, 109, 113, 114, 115, 119, 120, 125, 126, 127, 128, 129, 131, 132, 133, 137, 103, 150
"anomalous" phyllotaxis, 17, 49
Archimedean phyllotaxis, 28, 40, 44, 69, 70, 71, 74, 77, 119, 120
cone phyllotaxis, 89
cylindrical phyllotaxis, 15, 21, 23, 79, 80, 81, 82, 88, 90, 91, 92, 93, 95, 99, 102, 104, 108, 127
Fibonacci phyllotaxis, 14, 18, 25, 29, 35, 47, 52, 69, 70, 81, 82, 85, 91, 94, 95, 106, 119, 150
logarithmic phyllotaxis, 74, 75, 76, 78
multi-pair phyllotaxis, 60, 61, 93, 100
non-Archimedean phyllotaxis, 69, 70
planar phyllotaxis, 15, 21, 81, 91, 95, 97
power phyllotaxis, 70, 71, 72, 73, 74
phyllotaxis rises, 14, 35, 38, 68, 71, 75, 77, 84, 85, 89, 92, 96, 105, 131, 150
Plastochrone ratio R, 23, 27, 28, 74, 76, 107, 125, 128, 132
primordium, 13, 18, 20, 21, 23, 24, 40, 68, 105, 107, 109, 110, 111, 112, 113, 114, 121
proprimordium, 110, 111

R
recursive sequence, 5, 6, 7, 8, 16, 18, 58, 68, 106, 144
Fibonacci-like sequence, 8, 17, 45, 46, 47, 48, 54
generalized recursive sequence, 2, 6, 7, 8, 14, 18, 59, 60, 61, 64, 65, 66, 92, 95, 96, 104, 105, 106, 108, 131, 132, 143
non-multiple recurrent sequence, 17, 49, 50, 51, 52, 53, 54, 57, 59, 65, 90, 91, 92, 96, 131
RH-ratio, 84, 85, 86, 87, 88, 91, 92, 95
R-transformation, 121, 122

S

self-similarity, 44, 74, 88
spiral, 1, 2, 9, 10, 11, 13, 14, 16, 18, 19, 21, 22, 23, 24, 27, 29, 31, 32, 34, 41, 42, 45, 47, 48, 51, 52, 53, 60, 61, 62, 64, 67, 68, 69, 70, 72, 74, 76, 77, 78, 97, 98, 99, 102, 107, 108, 115, 118, 121, 125, 128, 131, 132, 133
    Archimedean spiral, 10, 11, 27, 43, 44, 67, 68, 69, 72, 77, 81, 97, 104, 158
    cylindrical spiral. *See* helix
    genetic spiral, 21, 22, 23, 27, 29, 30, 31, 42, 45, 47, 50, 51, 52, 59, 60, 61, 62, 69, 70, 73, 74, 76, 77, 83, 95, 97, 98, 99, 101, 107, 108, 118, 125, 128, 132
    logarithmic spiral, 10, 11, 75, 76, 77, 78, 158
    parabolic spiral, 10, 72
    planar spiral, 9, 10, 108, 121
    power spiral, 10, 70, 73, 76, 77, 78, 132, 158

V

visual effect, 18, 19, 28, 38, 75, 77, 85, 89, 105, 131
visual perception, 18, 19, 33, 36, 39, 61, 107, 125, 131

Y

Y-funnel, 122

ABOUT AUTHOR

Boris Rozin was born in Kiev, Ukraine. In 1981, he graduated from the math and physics High School #145 in Kiev and could not enter the top universities in Moscow and Kiev due to governmental anti-Semitism in the USSR. In 1989, he graduated with an Honors degree from the National Vinnitsa Technical University in Ukraine with a master's degree in Computer Science and Engineering. Immediately after graduation, he started working as a full-time assistant professor at the Department of Applied Mathematics and Computer Science of the National Vinnitsa Technical University. He taught: Applied Theory of Digital Devices, Mathematical Modeling, Discrete Mathematics and Graph Theory. Along with teaching, he studied in the PhD program, but due to anti-Semitism, the university administration put obstacles in the way of getting the degree.

In 2001 Boris received the refugee status in the US as being nationally persecuted and immigrated to the USA.

He now is an independent researcher and resides in Columbia, SC with his beloved family.

The author is open to partnership with researchers and research organizations.
borisrozin64@gmail.com
https://www.researchgate.net/profile/Boris_Rozin2
YouTube channel **Double Helix of Phyllotaxis**

CPSIA information can be obtained
at www.ICGtesting.com
Printed in the USA
LVHW080757050221
678279LV00033B/13